Transnationalism, diaspora and migrants from the former Yugoslavia in Britain

The geopolitical area of what once constituted Yugoslavia has been a region of significant migration since the 1960s. More recently, the conflicts in the region were the catalysts for massive displacements of individuals, families and whole communities. Thus far, there has been a gap in the literature on the qualitative experience of migrants from the former Yugoslavia through the twin theoretical lenses of transnationalism and diaspora.

This book offers an ethnographic account of migration and life in the diaspora of migrants originating from the former Yugoslavia and now living in Britain. Concepts such as the development of cultural beacons and diasporic borrowing are introduced through the ways in which migrants from the region form community associations and articulate – or avoid – such affiliations. The study examines the ways in which the experience of migration can be shaped by the socio-political contexts of departure and arrival, and considers how the lexicon associated with the act of migration can weave itself into the identities of migrants. The ways in which the trans-national and diasporic spaces are dictated by certain narratives, for example the allegory of dreaming and the language of guilt, are explored. It also investigates migrants' ongoing connection with the homeland, considering social and cultural elements, their reception in the UK and British media representations of Yugoslavia.

Contributing to the knowledge on the experiences of migrants from a part of the world which has been under-researched in terms of its migrating populations, this book will be of interest to students and scholars of Political Geography, Social Geography, Eastern European Politics, and Migration and Diaspora studies.

Gayle Munro has worked in the area of social research for over fifteen years for institutions both within the UK and internationally. She currently manages the research team at a London-based international charity where she conducts research with vulnerable and marginalised groups and on social exclusion.

Transnationalism, diaspora and migrants from the former Yugoslavia in Britain

Gayle Munro

Routledge
Taylor & Francis Group

LONDON AND NEW YORK

First published 2017
by Routledge

2 Park Square, Milton Park, Abingdon, Oxfordshire OX14 4RN
52 Vanderbilt Avenue, New York, NY 10017

Routledge is an imprint of the Taylor & Francis Group, an informa business

First issued in paperback 2020

British Library Cataloguing in Publication Data
A catalogue record for this book is available from the British Library

Library of Congress Cataloging in Publication Data
A catalog record for this book has been requested

ISBN: 978-1-138-69778-2 (hbk)
ISBN: 978-0-367-67052-8 (pbk)

Typeset in Times New Roman
by Deanta Global Publishing Services, Chennai, India

For my parents

The War of the Roses

Once I had a garden of roses.
I used to celebrate their smell and worship their forms.
Many passers-by would stop on their way and bend their heads
    to their beauty.
Until one night a rock-slide destroyed my garden, mashing
    rose petals with the soil.

On the ruins of their smells I have built again the garden
    and enclosed it with a wall.
Passers-by said –
We will help you to build a stronger and a taller wall.
And they built it. And they set night-watches to protect my
    garden from the rocks, and other misfortunes.
Now many more strangers were coming from everywhere to
    offer admiration to this Kingdom of Beauty.
We want to look after them too, they said to the
    night-watchers.
We want to be close to them and drink from their petals.
No, said the night-watchers, we are the only keepers of the
    Beauty.
And the others said, they belong to us as much as to you.
Why are you hiding them from us?
That is how the war of the roses started.
In anger and hatred they destroyed my garden.
So I began to cultivate the garden in myself.
The garden of the imprinted and colours of my roses.
Since then, I have met many others with gardens inside
    themselves.

Edin Suljić, 1994

Contents

Acknowledgements   viii
Abbreviations   x

Introduction   1

1 Contexts of departure   13

2 Contexts of arrival and reception   31

3 The lexicon of the migration experience   57

4 Intangible transnationalisms: The allegory of dreams   79

5 Cultural banks and beacons   93

Conclusion   107
Appendix   115
Index   119

Acknowledgements

Much of this book is based on doctoral work carried out at University College London. I would like to express grateful thanks and appreciation to my supervisor Claire Dwyer for her encouragement, support and patience. I was extremely fortunate to have, in Claire, a supervisor who has provided helpful, constructive, timely and honest advice over a number of years, for which I am extremely grateful. I would also like to thank other supervisors, Fiona Adamson and Ben Page for their input into the project at different points and Alan Latham and Gregory Kent who examined the thesis. Particular thanks are due to Greg for his encouragement and interest in the project.

I acknowledge the help and assistance of staff at the libraries and archives of University College London, the School of Slavonic and East European Studies, the Museum of London and the Imperial War Museum. I am very grateful for the assistance of the staff at Routledge and Deanta Publishing Services, especially Faye Leerink, Priscilla Corbett Sarah Douglas, Rebecca Lawrence and Michael Paye for their work on the manuscript, their help and support and for being so quick to respond to any queries I had. I also appreciated the feedback and comments from the two anonymous reviewers.

A number of authors and publishers kindly granted copyright permissions for the inclusion of extracts from their work. Thank you so much to Edin Suljić for his permission to cite his lovely poem in full. Grateful thanks are due to: Hearing Eye, London for permission to include extracts from *The Flying Bosnian: Poems from Limbo* by Miroslav Jančić (1935–2004); Vesna Goldsworthy and Atlantic Books for permission to cite from *Chernobyl Strawberries*; and Granta in the UK and Counterpoint in the US for permissions for *Bluebird* by Vesna Marić. Extracts from *Bluebird* (2009) are reprinted by permission of Counterpoint. Credits are due as follows regarding extracts from Eva Hoffman's *Lost in Translation* (1989): For the UK, published by Willian Heinemann, reproduced by permission of The Random House Group Ltd; for the US, used by permission of Dutton, an imprint of

Penguin Publishing Group, a division of Penguin Random House LLC. I am grateful to the Museum of London for permission to include extracts from interviews recorded for the Evelyn Oldfield Unit Refugee Communities History Project, archived at the Museum of London.

This project would not have been possible without the contribution of its participants whose voices, I hope, I have represented as faithfully as possible. I am so grateful to those who shared aspects of their lives with a stranger and who willingly offered pieces of their histories without knowing how such personal testimonies would ultimately be treated. I would like to express sincere thanks therefore to all respondents, interviewees and other participants who remain anonymous to the reader but whose stories are living testimonies of the challenges faced when making lives (willingly or otherwise) in a country other than the one of birth.

The most heartfelt thanks are due to my family. I have been very fortunate to have such supportive parents who have always stressed the value of learning and education as something to be cherished and enjoyed for its own sake. My mum and dad have always been enthusiastic and encouraging of my education and other pursuits – sometimes even when they may have found my particular passions quite bewildering. Posthumous thanks are given to my grandparents who always showed an interest and offered gentle advice and support. I would like to thank my children, Mia and Noam – this project is older than them both and they have been very patient in accommodating my 'homework'. Finally, I thank my partner Ervin, who has made a journey of his own, for listening, for his strength and resilience and for his forbearance in putting up with me on top of everything else. His contribution to this project has run as a constant thread throughout and as such, is immeasurable.

Abbreviations

EDM	Early Day Motion
EU	European Union
EVW	European Voluntary Worker
FRY	Federal Republic of Yugoslavia
GCSE	General Certificate of Secondary Education
KLA	Kosovo Liberation Army
ICT	Information, Communications and Technology
ICTY	International Criminal Tribunal for the former Yugoslavia
LSE	London School of Economics
MP	Member of Parliament
NATO	North Atlantic Treaty Organisation
NGO	Non-Governmental Organisation
PR	Public Relations
PTSD	Post-Traumatic Stress Disorder
RGS	Royal Geographical Society
SFRY	Socialist Federal Republic of Yugoslavia
SRF	Serbian Relief Fund
SSEES	School of Slavonic & East European Studies
UN	United Nations
UNHCR	United Nations High Commissioner for Refugees
VRS	Vojska Republike Srspke (Bosnian Serb Army)
WWI	World War One
WWII	World War Two

Introduction

> I sit in my garden now looking at the roses, it's a lovely evening isn't it? These are proper roses. Before I left Sarajevo, there were lots of 'roses' down our road, what was left of it. That's what I have now – memories of those roses and real ones growing in my new garden. New roses. That's what I try to hang on to these days. But the old ones will always be there – in my mind – even if they are trying to pour fresh concrete all over them.

The words above were spoken by Bosnian migrant Sara, in the context of a conversation about her move to Britain and about her life now in London compared with her life then in Sarajevo. Sara, in comparing the roses in her garden with Sarajevo roses, is referring to the craters in the roads and pavements in the Bosnian capital made by mortar explosions during the siege of Sarajevo by Bosnian-Serb forces, which were subsequently filled in with red resin, making patterns in the asphalt forming 'flower-like' shapes. As reconstruction continues in Sarajevo, these 'roses' – reminders of those killed or injured on a daily basis on the streets of the Bosnian capital during the siege – are slowly disappearing. Those roses have not disappeared however in the memories of Sara and others like her who have made their lives in Britain, some of whom may enjoy the new roses in their new gardens, but who carry with them the legacies of that conflict.

Whilst the majority of those living in present-day Britain who were born in what was Yugoslavia came as a direct or indirect result of the violent conflicts in the region in the 1990s and events associated with the disintegration of the Socialist Federal Republic of Yugoslavia, that period was by no means the beginning of migration from the region to Britain. WWI and WWII had also seen significant waves of migration from the region across Europe and beyond, with some of those who left, Serbs in particular, coming to Britain at that time. The geopolitical area of what once constituted Yugoslavia was then a region of significant outmigration since the 1960s. More recently, the conflicts in the territories of the former Yugoslavia were the catalysts for massive displacements of individuals, families and whole communities to neighbouring countries, across Europe and beyond. Some of those who have left have since returned to homes changed significantly by the genocidal policies of 'ethnic cleansing'. Others have found their 'durable solutions' not in the original homeland but within the diaspora. Still others have migrated from areas not immediately affected by violence or the threat of violence but have made the decision to establish homes elsewhere as a result of other personal factors, in some cases unrelated to the conflicts.

Despite the scale of migration out of the former Yugoslavia, there has been little literature on the nuanced experiences of those from the region who have left their homes behind to seek new lives elsewhere. Diaspora studies on migrants from the former Yugoslavia have tended to focus on the experiences of migrants from 'one country of origin in multiple host environments' approach (Franz, 2005; Čolić-Peisker, 2008) or have articulated the voices of migrants of one ethnicity within one host state (Coughlan & Owens-Manley, 2006; Procter, 2000). Less covered in the literature has been an examination of the stories of those from different parts of what was Yugoslavia within one host state, especially through the lens of not just ethnicity, but also exploring other potential variables which could impact upon the experience of migration. This book makes a contribution therefore to the knowledge on the diasporic and transnational experiences of migrants from a part of the world which, whilst having held the focus of political scientists and conflict specialists, has been under-researched in terms of its migrating populations and diasporas.

Scholars of migration within different disciplines have historically tried to understand and conceptualise the phenomenon of the movement of peoples from one bounded area of residence to another. The experience of migration as an area of study has attracted the interest of geographers, anthropologists, political scientists, economists, sociologists, demographers, statisticians, artists, writers and film-makers, who will all approach the phenomenon with a different angle and lens of focus. The present time sees this focus as more intense than ever with circumstances facing those refugees fleeing the conflict in Syria and other states becoming increasingly desperate. European states argue about quotas, policy makers struggle to respond coherently and in a timely fashion to rapidly-changing events, media headlines inflate and conflate the discourse and academics hold intense debates over the word 'crisis'. In the meantime, conditions within refugee camps further deteriorate and generations of refugees are left to attempt to make an existence for themselves and their families.

As the movement of people from one bounded territory to another escalates and developments in ICT allow international migrants to maintain a different level and intensity of relationship with each other and with those in the country of origin than might have previously been possible, migration scholars have sought to develop new understandings of the ways in which to capture these flows and ties between the migrant in his/her 'new home' and those left behind. Prompted by the popular discourse surrounding the ongoing refuge and migration crisis, the language associated with migration has been thrown into sharper focus, particularly related to the motivation for migration. Much has made of the recent discourse of (mis)understandings of labelling 'refugees' as 'migrants' and the conflation of (economic) migration with asylum, or 'voluntary' with 'forced' migration. Research for this project with those from the former Yugoslavia who have made their homes in Britain has shown that the decision-making process is complex and can rarely be attributed to any one motive or purpose. In settling on one term which could capture a range of migration experiences, my presentation of a 'migrant' in the context of this project is someone who has made his/ her home – either permanently or temporarily – in a country other than the one in which he/she was born. I discuss later (Chapter 3) how the lexicon surrounding the migration experience, and in particular relating to immigration status, was understood by the participants in my research. However, by using the term 'migrant' I am making no value judgement on the motivation behind the decision for leaving the country of origin and moving to Britain.

Conceptual understandings

This book was born out of a desire to explore the relationship that a migrant forms and develops with his/her country of origin, the 'homeland'. What are the factors of influence which can shape and form any changes in that relationship? What are the variables which determine whether a migrant maintains a relationship with his/her country of origin or not? What might such a relationship look like and how is it connected to existing or developing ties with the 'new' home? The following extract from Eva Hoffman's *Lost in Translation* encapsulates how the condition of being in-between two worlds can colour the experience of migration:

> I can't afford to look back, and I can't figure out how to look forward […] Time has no dimension, no extension backward or forward. I arrest the past and I hold myself stiffly against the future; I want to stop the flow. As a punishment, I exist in the stasis of a permanent present, that other side of 'living in the present', which is not eternity but a prison. I can't throw a bridge between the present and the past, and therefore I can't make time move.
>
> (Hoffman, 1998)[1]

The concept of transnationalism, if interpreted qualitatively, can capture such a condition of 'in-betweenness' as articulated in the quote above. I use a lens of focus rooted in the concepts of transnationalism and diaspora as paradigms through which non-linear migrant flows and spaces in which ties with both the 'homeland' and the 'new home' can be expressed and flourish (or not). My research questions were broadly framed around exploring how migrants coming from the former Yugoslavia to Britain exhibit transnational and/or diasporic identities and relationships, or how do they express transnational and/or diasporic *being and belonging*. And what are some of the variables which may impact upon such transnational and/or diasporic affiliations, or the development of transnational and/or diasporic space?

Researchers of migration have adopted different understandings of what could be termed transnational migration and the individual actors within – transmigrants – often depending on whether or not they view transnationalism as a different conceptualisation of the migration experience from previous, already established frameworks. Introduced into the main lexicon of migration scholarship in the 1990s by the work of Glick Schiller *et al.* (1992a, 1992b *inter alia*) and Basch *et al.* (1994), and then later developed by Portes *et al.* (1999 *inter alia*), transnationalism as a concept broadly refers to 'a social process in which migrants establish social fields that cross geographic, cultural, and political borders' (Glick Schiller *et al.*, 1992b: ix). Distinguishing between 'internationalism' (referring to interactions between national governments, diplomacy, formal trade) and 'transnationalism', Vertovec (2009: 3) conceptualises transnational practices as 'sustained linkages and ongoing exchanges among non-state actors based across national borders'. Some scholars disagree over the extent to which transnationalism as a concept can be considered a departure from already established ways of viewing the migration experience (Waldinger & Fitzgerald, 2004; Foner, 1997; Morawska, 2001). Others highlight how a focus on the importance of global capital forces in the development of transnational ties has possibly led to a neglect of the transnational lens of those who may not have left their home country voluntarily (Al-Ali & Koser, 2002) and also to an over-emphasis on the economic variables of any transnational activities (Vertovec, 2001). Portes *et al.* (1999) suggest prescriptive conditions under which individuals and families should be operating before a migrant can be considered to be a transnational migrant. Critics of such an approach have rejected the application of such restrictive terms, particularly those who are working within a European context (Jackson *et al.*, 2004; Al-Ali & Koser, 2002). Studies of transnationalism often begin with situating the context by highlighting the US-bias in the field but this has been changing more recently as the field develops.[2]

If the 'transnational turn' in migration studies can be considered to be a fairly recent development, then the study of migration through a diasporic lens is situated across significantly more well-trodden ground. Most discussions on diaspora will preface any theoretical or empirical commentary with a review of 'classical' diasporas before continuing with examinations of diaspora in the present day (Braziel & Mannur, 2003; Reis, 2004). Scholars of diaspora have highlighted the dilution of the term, which through everyday usage and adoption in journalistic contexts, has led to what diaspora theorists have lamented as misunderstandings and misapplications of the concept (Cohen, 1997; Sheffer, 2003; Brubaker, 2005; Tölölyan, 2010; Kokot *et al.*, 2004). One of the hallmarks of diaspora as a social form as presented in the literature is the 'triadic relationship' (Sheffer 1986) between: a) globally dispersed yet collectively self-identified ethnic groups; b) the territorial states and contexts where such groups reside; and c) their homeland states and contexts (Vertovec 1999: 450). Safran (1991) set out six characteristics of what he considered to constitute present-day diasporas; however, Reis (2004) observes how unusual it would be to identify all six in the makeup of present-day diasporas. In terms of migrants from the former Yugoslavia in Britain, the lack of emphasis on the homeland as the focal point of return in Sheffer's (2003) definition is to be welcomed; however, the 'showing solidarity with their group and their entire nation' element is still problematic to apply to migrants from a region where nation, nationality, ethnicity and religion are all conflated terms, associations which may have contributed to the migrant's forced exile from the homeland.

In lamenting the dilution of the term 'diaspora' as it gains more popular currency, the voices of those migrants who could be considered to be diasporic actors are sometimes missing from the debate. The majority of my respondents may not have used the term 'transnational' to apply to their experiences but all referred to being 'in the diaspora' in very general, matter-of-fact ways which may not fit into the standard academic definition and, indeed if used in an academic context, would usually risk being described as 'essentialist'. Through the experiences of my participants, I understand the development of the transnational space as an answer to the feeling of not being 'entirely there' or 'entirely here' but somewhere in-between.[3] The inhabiting of such a space will be qualitatively defined by a whole host of factors, such as the experience of migration, age, gender, place on the life-cycle, and so on.

My understanding of transnationalism differs from those who emphasise the group dynamic of transnationalism (Djelić & Quack, 2010) in favour of a more individualistic focus. A migrant can exhibit transnational behaviour without necessarily prescribing to the dominant socio-political discourse of the country of origin; indeed, transnational expressions are sometimes rooted in a rejection of such discourse. Diaspora, conversely, through the definitions in the literature, implies a focus on the collective – a common or shared sense of identity or experience. The 'shared collective memory' (Saffran, 1991; Lacroix & Fiddian-Qasmiyer, 2013), so crucial to many understandings of diaspora, may also be difficult to apply to entire 'nations' within the context of the former Yugoslavia but Halilovich (2011) has demonstrated how, in the Bosnian case, trans-*local* (as opposed to trans*national*) ties to the *zavičaj* within the Bosnian diaspora have been developed as a response to the genocidal practice of 'ethnic cleansing' (Halilovich, 2013: 231).

Terminology

It would be easy – without digesting the arguments of Glick Schiller, Portes and colleagues – to misinterpret 'transnationalism' for 'long-distance nationalism' – a subject which has already been investigated within a journalistic (Hockenos, 2003) and an academic context (Pryke, 2003; Skrbiš, 1999) in reference to the former Yugoslavia. Such studies have focussed on the activities of the different diasporas in relation to contributions – financial and others less benign in nature – to the conflicts in the region in the 1990s and in post-conflict reconstruction.[4] In light of the fact that many of those from the former Yugoslavia resident in Britain today are here directly or indirectly because of the *nationalist* policies and acts of *nationalist* political elites in the region, the use of the term trans*national* is understandably sensitive and could potentially carry with it negative and distressing connotations. It is partly to avoid such connotations that I avoided the use of the term transnational in my call for research participants or in framing the investigation for potential respondents. I also made no direct reference to the term diaspora as I wanted to avoid any researcher-led directions as to how migrants would define their position(s) and affiliation(s). I discovered during the course of the research how the labelling ascribed to certain processes can have an impact upon the way the project might be understood by potential participants.

   The case of the former Yugoslavia is a particularly interesting and challenging context in which to frame a debate about transnational migration or diasporic allegiances given the complexities of potential national, ethnic, religious, local loyalties (or lack thereof). The former Yugoslavia as a case-study offers the opportunity to make a contribution towards addressing the gap in the literature on transnationalism of studies which explore the experiences of refugees through the transnational lens (Al-Ali *et al.*, 2001), and the call for more comparative studies (Ostergaard-Nielsen, 2003a), particularly within a European context. A focus on migrants from different parts of the region therefore allows for the development of a more nuanced understanding of the paradigm of transnationalism rooted in the diversity itself of the myriad experiences of migration from the region. Even considering Britain as the country of destination is also potentially challenging with the union of England, Scotland, Wales and Northern Ireland. There were no participants in my project from Northern Ireland which is why this book refers to migrants to Britain as opposed to the United Kingdom. I refer throughout the book to the former Yugoslavia and the call for participants was also framed in that way. However, I had so few responses from migrants from Macedonia and Montenegro that I made the decision to only focus on the reported experiences of those migrants who engaged in greater numbers, namely, respondents from Bosnia and Herzegovina (referred to hereafter as Bosnia), Croatia, Kosovo, Serbia and Slovenia. In making general discussion on the ethnicity of respondents, I refer to Croats/Serbs/Slovenes when referring to ethnicity and Croatians/Serbians/Slovenians when discussing questions of nationality or citizenship. I respect throughout the individual's self-identifier when referring to the ethnicity of a respondent. When referring to ethnicity of respondents from Bosnia, I make reference to Bosnian Croats, Bosnian Serbs and Bosnians. Implicit in the term 'Bosnians' I am referring to Bosnians of Muslim ethnicity. In cases where I am discussing those migrants from Bosnia as encompassing those of Croat, Muslim and Serb ethnicity as a collective, I make reference to 'migrants from Bosnia'. Regarding respondents from Kosovo, I refer to Kosovo with the '-o' as opposed to the '-a' ending as Kosovo is the internationally recognised name for the state. I have used the English spelling of chetnik as opposed to *četnik* mainly because the plural chetniks is more commonly understood than *četnici*.

Research design and chapter structure

The methods I used throughout the research reflect my attempts to capture the 'holistic' nature of the transnational experience. I discuss this in more detail in the Appendix; however, I employed a constructivist-grounded approach to the project (Glaser & Strauss, 1967). The methods used involved: a survey of 179 migrants from the former Yugoslavia resident in Britain; interviews with 46 individuals living in Britain at the time of the interview; discourse analysis of archival material, media and parliamentary material, autobiographical contemporary fiction, material in cyberspace and ethnographic observation at events held (mainly in London) related to the former Yugoslavia. Throughout the project my focus has been on what could be considered as the articulation of transnational practices and activities (or their absence) and transnationalism as expressed through less quantifiable means – transnational identities, affiliation and those expressions of transnationality which are more challenging to capture in a measurable way.

Chapter 1 provides the contexts in which migrants from the former Yugoslavia made their decisions to leave the 'homeland' and make the journey to Britain – to make either temporary or permanent new homes. Although the conflicts in the territories of the former Yugoslavia were the main catalysts for migration from the region to Britain, the violence or threat of violence in the 1990s was not the only driver of migration experienced by participants in the research study on which this book is based. This chapter therefore provides some historical context to the migration experience as articulated by my respondents by outlining some of the socio-political contexts in the former Yugoslavia which had the potential to generate outward migration before focusing on the personal motivations for migrating as narrated by participants in the research. The last part of the chapter provides an analysis of the experience of departure itself as a catalyst for, in some cases, an intense and complex ongoing relationship with 'home'.

Literature in the context of reception (Portes & Rumbaut, 2006) delineates the ways in which the experience of migration and more crucially the relationship with the country of origin can be shaped and influenced by the social, economic, political and cultural environments into which the newcomer arrives. Chapter 2 therefore discusses this 'context of reception' for migrants from the former Yugoslavia arriving in Britain. The British government's position on migration from the region has changed significantly over the years and the legislative and policy frameworks which migrants from the former Yugoslavia would have been negotiating would depend on from where and when they made their journeys. The immigration system itself therefore is considered as a key player in the formation of transnational and diasporic relationships. This chapter then moves on to a discussion of how popular and more elite representations of the former Yugoslavia may have had an impact upon the context of reception for migrants arriving at different times and finishes with a discussion of how certain events triggered a reaction on the part of some migrants from the region in Britain – both on an individual and a community level.

Chapter 3 starts by considering the ways in which those from the former Yugoslavia articulate their identities as migrants. How are the vocabulary of migration and certain 'labels' such as 'refugee', 'exile', 'migrant' understood and (re)interpreted over time and space? How are the experiences of co-migrants in some cases appropriated by others and rearticulated to fit the dominant diasporic narrative? The chapter then moves on to a discussion of the ways in which migrants identify (or not) on different levels with the region of origin. This chapter concludes with a discussion of the impact that the emotion of guilt, which featured as an ever-present motif throughout the respondent narratives, can have upon transnational and diasporic identities and relationships.

The allegory of dreams is one which runs through much of the discourse on migration, both in academic and more popular narratives. The focus is often on unrealised dreams, the fracture between imaginings of a better life elsewhere and the reality of the migration experience. Chapter 4 discusses how the dream allegory plays out in memories of the homeland and continued identity formations. This narrative is dominated not by the aspirational 'hopes and dreams' motif, so often presented within the discourse on migration; rather, through the dreams of my respondents it is possible to discern the echoes of their lives in the homeland and the blueprints of lives running in parallel.

Chapter 5 argues that diasporic communities, as a means of reinforcing their collective identities, form 'banks' of cultural references which other members of the diaspora can then draw upon in the formation of their own transnational affiliations. These banks can then be – especially when they are formed in conjunction with or targeted at the 'host' environment – used as agents to effect a desired change. Any reading of the development of diasporic space formed by those from the former Yugoslavia would need to be considered against the backdrop of the practice of cultural genocide which characterised the conflicts of the 1990s (Walasek *et al.*, 2015). This chapter conceptualises the construction of community spaces within the diaspora as 'cultural beacons' – sites which can act as a welcome to those who similarly identify and as a deterrent to the uninvited or to those who reject the connotations of such spaces.

In all considerations of articulations of transnationalisms of different types, I have been mindful of what has been termed as 'sampling on the dependent variable' in the study of transnational case-studies (Faist, 2010). I also pay due consideration to those experiences which could not be considered to be transnational in nature – irrespective of how broadly that concept is understood. I have also been conscious of the cautions against focussing on ethnicity to the exclusion of other possible variables which could influence transnational behaviour (Al-Ali *et al.*, 2001). As Foner (1997: 370) has stated: 'some groups [and places] are likely to be more transnational than others – and we need research that explores and explains the differences'. In attempting to explore the transnational character of migration in the context of movement from the former Yugoslavia to Britain, I have therefore sought to focus on socio-demographic variables other than a sole emphasis on ethnicity. I consider ethnicity alongside potentially influential variables such as gender, age, family demographic, immigration status, motivation for migrating and time of migration.

Notes

1 Eva Hoffman (1998) writing after her family moved from her native Krakow to Vancouver in 1959.
2 For a fuller review of the literature see Bauböck and Faist (2010) and Vertovec (2009). For a summary review and annotated bibliography see Munro (2015).
3 Such a state of being 'neither here nor there' has been articulated by others, for example Al-Ali and Koser (2002).
4 'Kosovo: Overseas Albanians send recruits, arms and money', Transnational Communities Programme: www.transcomm.ox.ac.uk/traces/iss2pg2.htm (accessed 22 April 2016) and 'Kosovo: The KLA and the Albanian diaspora', Transnational Communities Programme: www.transcomm.ox.ac.uk/traces/iss3pg2.htm (accessed 22 April 2016).

References

Al-Ali, N. and Koser, K. (2002) 'Transnationalism, international migration and home', in N. Al-Ali and K. Koser (eds) *New Approaches to Migration: Transnational Communities and the Transformation of Home*. London: Routledge.

Al-Ali, N., Black, R. and Koser, K. (2001) 'Refugees and transnationalism: The experience of Bosnians and Eritreans in Europe', *Journal of Ethnic and Migration Studies* 27(4): 615–634.

Basch, L., Glick Schiller, N. and Szanton-Blanc, C. (1994) *Nations Unbound: Transnational Projects, Postcolonial Predicaments, and Deterritorialized Nation-States*. New York: Gordon and Breach.

Bauböck, R. and Faist, T. (eds) (2010) *Diaspora and Transnationalism: Concepts, Theories and Methods*. Amsterdam: Amsterdam University Press.

Braziel, J.E. and Mannur, A. (eds) (2003) *Theorizing Diaspora: A Reader*. Oxford: Blackwell Publishing.

Brubaker, R. (2005) 'The "diaspora" diaspora', *Ethnic and Racial Studies* 28(1): 1–19.

Cohen, R. (1997) *Global Diasporas: An Introduction*. London: University College Press.

Čolić-Peisker, V. (2008) *Migration, Class and Transnational Identities: Croatians in Australia and America*. Urbana: Illinois University Press.

Coughlan, R. and Owens-Manley, J. (2006) *Bosnian Refugees in America: New Communities, New Cultures*. New York: Springer.

Djelić, M.L. and Quack, S. (2010) *Transnational Communities: Shaping Global Economic Governance*. Cambridge: Cambridge University Press.

Faist, T. (2010) 'Diaspora and transnationalism: What kind of dance partners?', in R. Bauböck and T. Faist (eds) *Diaspora and Transnationalism: Concepts, Theories and Methods*. Amsterdam: Amsterdam University Press.

Foner, N. (1997) 'What's new about transnationalism? New York immigrants today and at the turn of the century', *Diaspora* 6(3): 355–75.

Franz, B. (2005) *Uprooted and Unwanted: Bosnian Refugees in Austria and the United States*. College Station: Texas A&M University Press.

Glaser, B. and Strauss, A. (1967) *The Discovery of Grounded Theory: Strategies for Qualitative Research*. Chicago, IL: Aldine.

Glick Schiller, N., Basch, L. and Szanton Blanc, C. (eds) (1992a) *Towards a Transnational Perspective on Migration: Race, Culture, Ethnicity and Nationalism Reconsidered*. New York: New York Academy of Sciences.

Glick Schiller, N., Basch, L. and Szanton Blanc, C. (1992b) 'Towards a definition of transnationalism: Introductory remarks and research questions', *Annals of the New York Academy of Sciences* 645: ix–xiv.

Halilovich, H. (2011) '(Per)forming "Trans-local" homes: Bosnian diaspora in Australia', in M. Valenta and S. Ramet (eds) *The Bosnian Diaspora: Integration in Transnational Communities*. Surrey: Ashgate Publishing.

Halilovich, H. (2013) *Places of Pain. Forced Displacement, Popular Memory and Trans-Local Identities in Bosnian War-Torn Communities*. New York: Berghahn.

Hockenos, P. (2003) *Homeland Calling: Exile Patriotism and the Balkan Wars*. Ithaca: Cornell University Press.

Hoffman, E. (1998) *Lost in Translation*. London: Vintage Books.

Jackson, P., Crang, P. and Dwyer, C. (eds) (2004) *Transnational Spaces*. London: Routledge.

Kokot, W., Tölöyan, K and Alfonso, C. (eds) (2004) *Diaspora, Identity and Religion: New Directions in Theory and Research*. London: Routledge.

Lacroix, T. and Fiddian-Qasmiyeh, E. (2013) 'Refugee and diaspora memories: The politics of remembering and forgetting', *Journal of Intercultural Studies* 34(6): 684–696.

Morawska, E. (2001) 'Immigrants, transnationalism, and ethnicization: A comparison of this great wave and the last', in G. Gerstle and J.H. Mollenkopf (eds) *E Pluribus Unum? Contemporary and Historical Perspectives on Immigrant Political Incorporation*. New York: Russell Sage.

Munro, G. (2015) 'Transnationalism. A review of the literature', *Studies on National Movements* 3.

Ostergaard-Nielsen, E. (ed.) (2003a) *International Migration and Sending Countries: Perceptions, Policies and Transnational Relations*. London: Palgrave Macmillan.

Portes, A. and Rumbaut, R. (2006) *Immigrant America: A Portrait*. Berkeley: University of California Press.

Portes, A., Guarnizo, L.E. and Landolt, P. (1999) 'The study of transnationalism: Pitfalls and promises of an emergent research field', *Ethnic and Racial Studies* 22(2): 217–237.

Procter, N. (2000) *Serbian Australians in the Shadow of the Balkan War*. Hants: Ashgate Publishing.

Pryke, S. (2003) 'British Serbs and long distance nationalism', *Ethnic and Racial Studies* 26(1): 152–172.

Reis, M. (2004) 'Theorizing diaspora: Perspectives on "classical" and "contemporary" diaspora', *International Migration* 42(2): 41–60.

Safran, W. (1991) 'Diasporas in modern societies: Myths of homeland and return', *Diaspora* 1(1): 83–99.

Sheffer, G. (ed.) (1986) *Modern Diasporas in International Politics*. New York: St Martin's Press.

Sheffer, G. (2003) *Diaspora Politics: At Home Abroad*. Cambridge: Cambridge University Press.

Skrbiš, Z. (1999) *Long-Distance Nationalism: Diasporas, Homelands and Identities*. Hants: Ashgate.

Tölölyan, K. (2010) 'Beyond the homeland: From exilic nationalism to diasporic transnationalism', in A. Gal, A.S. Leouissi and A.D. Smith (eds) (2010) *The Call of the Homeland: Diaspora Nationalisms, Past and Present*. Leiden: Brill.

Vertovec, S. (1999) 'Conceiving and researching transnationalism', *Ethnic and Racial Studies* 22(2): 447–462.

Vertovec, S. (2001) 'Transnationalism and identity', *Journal of Ethnic and Migration Studies* 27(4): 573–582.

Vertovec, S. (2009) *Transnationalism*. London: Routledge.

Vertovec, S. and Cohen, R. (eds) (1999) *Migration, Diasporas and Transnationalism*. Cheltenham: Edward Elgar.

Walasek, H., Carlton, R., Hadžimuhamedović, A., Perry, V. and Wik, T. (2015) *Bosnia and the Destruction of Cultural Heritage*. Surrey: Ashgate.

Waldinger, R. and Fitzgerald, D. (2004) 'Transnationalism in question', *American Journal of Sociology* 109(5): 1177–1195.

1 Contexts of departure

This chapter situates the contexts of departure in which migrants from the former Yugoslavia have made their decisions to make Britain their temporary or permanent home and discusses some of the multi-layered and mosaic-like motivations behind the migration decision. One of the aims of my research was to investigate the different conditions under which transnational expressions and affiliations of being and belonging can flourish (or not). The conditions in the country of origin prior to the point of initial migration and the context of departure are key variables to consider when exploring any transnational ties or affiliations. Those migrants from the former Yugoslavia who participated in my project arrived in Britain at different points over the period 1953–2010. My research participants demonstrated multiple and complex motivations for moving to and making their homes in Britain at different stages of their lifecycle and, depending on the period in which they made their journeys, have different experiences along and throughout their 'stories' of migration. The main aim of this chapter therefore is to begin to contextualise some of the varied experiences of the act of migration itself. It is beyond the scope of this book to provide an extensive history of such a complex region and that ground has been well covered elsewhere by others. I was more interested in exploring how the history of the region of origin was portrayed by my participants and how such portrayals have affected the transnational relationship. I therefore briefly refer to some key points as background which were foregrounded in the interviews and conversations with my research participants and signpost the reader who is looking for a more comprehensive historical narrative to other texts.

Leaving 'home': Migration from the former Yugoslavia

Migration scholars and particularly those with an interest in the former Yugoslavia have highlighted how the body of literature on migration from the region may not necessarily be as prolific as might be expected

(Kupiszewski cited in Valenta & Ramet, 2011: 1; Pryke, 2003). Whilst it is true that migration from the former Yugoslavia has probably not attracted as much attention as migration from other parts of the world, there have been a number of publications on the experiences of migrants from the region within the contexts of different countries of reception, including *inter alia*: Bosnians (Behloul, 2011; Coughlan and Owens-Manley, 2006; Eastmond, 1998; Franz, 2002 and 2005; Halilović-Pastuović, 2007; Halilovich, 2011a, 2011b and 2013; Jansen, 2008; Kelly, 2003 and 2004; Mišković, 2011; Slavnić, 2011; Wight, 2000); Bosnian Croats (Povrzanović Frykman, 2011); Croats (Carter, 2007; Čapo-Žmegač, 2007; Čolić-Peisker, 2008) Croats and Slovenes (Skrbiš, 1999); Kosovars (Dahinden, 2005 and 2009; Kostovicova Prestreshi, 2003); and Serbs (Lerch *et al.*, 2006; Procter, 2000; Pryke, 2003). Much of the literature considering the experiences of migrants from the region in Britain has been in the field of mental health and focuses on the experiences of refugees from Bosnia and Kosovo (Summerfield, 2003; Herlihy *et al.*, 2002; Fozdar, 2009). The discussion has often been presented with a policy focus either on health (Turner, 2003) or on the experiences of refugees in the context of discussion around British immigration policies (Balicki and Wells, 2005).

The region of Southeast Europe has historically been one which has demonstrated high population movements, even prior to the conflicts of the 1990s with labour migratory patterns to countries in Western Europe a particular feature of the demography of the region in the post-WWII era and especially in the 1960s (Baučić, 1974; Kosińksi, 1978). The tracing of migration patterns in the region has been an area of interest to local ethnographers as the region has undergone the redrawing of borders and territorial boundaries. Indeed, in more recent times, the work of entire institutions in the region has been devoted to attempting to trace the lineage of a particular ethnicity within an area – in an arena of territorial gains and losses, evidence of an established 'ethnic claim' of a disputed area can be crucial and subject to wide interpretation, manipulation and revisionism.

Halpern (1975: 77) observes that 'to understand fully the meaning of contemporary population movements a view of the past is essential'. This need for a historical context is no greater than across a geopolitical area that has been the site of considerable political and social flux for centuries, the effects of which are still being felt into the present day. The influence of myth, legend and historical story-telling on the identity-formation and collective consciousness of the various peoples of the region has been the subject of several studies both prior to and in light of the violent dissolution of the state of Yugoslavia (Ramet, 1996; Judah, 1997). The question of 'whose history' however is one which can be problematic in an area where versions of history have been moulded to fit the dominant political discourse and where the Orwellian 'who controls the past, controls the future' seems to sometimes be interpreted as an ideology to be aspired to.

In listening to the body of narratives of those who made their journeys from different parts of the former Yugoslavia, and in particular the ways in which stories were (re)interpreted and (re)told, it was possible to discern historical 'hooks' around which were formed individual and collective identities. The Battle of Kosovo at Kosovo Polje in 1389 was referred to, for example, by a number of respondents (Bieber, 2002) and also features in the discourse of British parliamentarians. Historical antagonisms between the different ethnic groups were referred to in vague detail by respondents from all parts of the region, especially men, of any age. It was from WWI onwards however where the history of the region seemed to have taken firmer root in the consciousness of my (particularly Serb) respondents. The beginning of WWI saw heavy losses on the part of the Serbian army, already depleted from the Balkan Wars just a few years previously. In October 1915 full-scale invasions on the part of German and Austro-Hungarian forces, combined with the Bulgarians entering the war forced a retreat of the Serbian army south through Montenegro into Albania and then by allied boats to Corfu and eventually Salonika. The conditions on the retreat and the suffering experienced by the Serbian army are well documented and have 'found a permanent place in popular historical memory' (Lampe, 1996: 104).

During the tumultuous events of WWI, we can see evidence of formalised and intense relationships between Serbia and Britain which have taken firm root in the collective consciousness of British Serbs. Accompanying the Serbian army on its retreat to Salonika was Flora Sandes, the only British woman to have officially served as a soldier on the front-line during the conflict, and a number of medical staff stationed in units established by British relief agencies in Serbia during the war (Miller, 2012). The Serbian Relief Fund (SRF) from its London base was operating to raise awareness and funds to come to the assistance of the plight of the Serbs. The SRF established a number of medical units within Serbia and provided funds for other missions sent from Britain, including the Scottish Women's Hospitals (Aldridge, 1916; Berry & Berry, 1916). The work of those British medical units and the sacrifice made by the doctors and nurses who lost their lives to typhus or other diseases is remembered in Serbia and by British Serbs up until the present day with significant transnational links remaining between medical communities in Britain and Serbia (Foster, 2013; Mikić, 2007; Liddington, 2011; Leneman, 1994). A number of Serb respondents made reference in their narratives to this period of history as evidence of the close relationship between Britain and Serbia – some in the context of disappointment expressed at later involvement by Britain in military action against the Serbs. British–Serb relations during WWI continue to be the focus of commemorations in Serbia to the present day (Miller, 2012; Mikić, 2007) as well as fundraising initiatives in Britain.

References to Britain's relationship with Yugoslavia in the years before and after WWII also featured in the personal narratives of migrants from the region to Britain and (as discussed in the following chapter) within British political discourse surrounding the more recent Yugoslav conflicts of the 1990s. Ties between the British monarchy and Serbian royal family and British allegiances at different times of the war were also referenced by Serb respondents.[1] It is well documented how nationalist elites within Yugoslavia drew upon historical and popular narratives of the extreme violence around the actions of chetniks and *ustaše* during WWII in their nationalist rhetoric and their manipulations of the emotions of the local populations, many of whom would have had family members or friends who had been affected by the violence, in the years leading up to and during the conflicts and the more recent conflicts of the 1990s (Silber & Little, 1996).

Following the war, a number of Serbs in prisoner of war camps in Italy and other parts of Europe were recruited for unskilled labour in Britain. The common perception of these post-war Serbs amongst my research participants is that they were loyal to Draža Mihailović[2] and the royalists and reflections of these affiliations can be seen in the names of some Serbian organisations in Britain today.[3] Bosnians who had been interned in camps in Austria, Italy and Germany were also recruited by the British labour commission (Fazlić, 2009).

The Socialist Federal Republic of Yugoslavia (SFRY) remained a union until 1991 under the slogan of 'Brotherhood and Unity' (*bratstvo i jedinstvo*), for the majority of that time under the leadership of Josip Broz Tito. Following a split with Stalin in 1948, Yugoslavia underwent a period of reforms and decentralisation of power. A version of market socialism was introduced with the relaxation of borders. Following the Croatian Spring of 1971 a new constitution was adopted giving greater autonomy to the republics, a move not welcomed by Serbia who viewed the reforms as concessions to Albanian nationalists in Kosovo. One characteristic of the regime was the agreements made with European governments, Germany in particular, to the sending of *gasterbeiter* or guest workers, seen as a means of boosting the Yugoslav economy whilst addressing the German labour shortage (Baletić, 1982).

Following Tito's death in 1980, Yugoslavia was fraught with undisguised ethnic politics, an unstable federal foundation, high levels of unemployment and an accumulation of economic problems.[4] A rising sense of nationalistic fervour within Serbia in the form of political discourse and popular culture was channelled through calls by Milošević to centralise power, details of which he propounded during a series of large public meetings throughout Serbia. Increasingly nationalist rhetoric was expressed in both Croatia and Serbia. References were made on television and in the press to the 'genocidal' threat to the Serb nation on the part of Albanians in Kosovo and Croatian '*ustaše*'. In Croatia in turn, people were once again encouraged to fear a chetnik force.

The short and relatively bloodless conflict on Slovenian soil (also known as the ten-day war) following Slovenia's secession from Yugoslavia was the beginning of years of the most traumatic conflicts seen in Europe since WWII. In Croatia, areas such as Krajina, Vukovar, Osijek and Dubrovnik were savaged with wide-spread atrocities committed and charges of war crimes. The first reports of armed conflict in Bosnia were received in April 1992, following a referendum on independence in February 1992. The exact numbers of those killed, missing and displaced of various ethnicities during the conflicts are the subject of much dispute, with those engaged in the difficult work of trying to gather evidence being in receipt of death threats themselves.[5]

Unlike in Croatia, which saw more localised destruction, Bosnia as a territory was utterly devastated. It is difficult to put into words the horror that was experienced by the local populations experiencing the realities of conflict on the ground. Tomes have been written on the conflicts in what was Yugoslavia and the purpose of this book is not to go over well-trodden ground on the causes of the conflict or to analyse in detail the horror of those years.[6] Others have already followed that path. But the majority of migrants from the region resident in Britain today have made their journeys – either directly or indirectly – because of the tragic results of the actions of ethno-nationalist aggressors, the lack of means of those targeted to defend themselves and the lack of will on the part of governments from outside the region to take action to intervene, either individually or as a collective.

The late 1990s and early 2000s were characterised in parts of the former Yugoslavia by continued ethnic and civil unrest, UN-imposed economic sanctions, military action by NATO, economic collapse and further nationalist rhetoric. Other parts enjoyed economic growth and a welcome by the European Union. Full-scale conflict was fought over Kosovo between the armed forces of the Federal Republic of Yugoslavia (FRY) and the Kosovo Liberation Army (KLA) between February 1998 and June 1999. However, the years leading up to full conflict and the recognition of Kosovo's independence were punctured with episodes of brutality, ethnic cleansing and violent reprisals.[7] Following the failure of diplomacy at Rambouillet in February 1999, as a means of forcing the withdrawal of Serb forces from Kosovo, NATO led a 78-day campaign of air-strikes against Serbia from March to June 1999. The subject of much controversy and debate as the strikes were carried out without the approval of the UN Security Council, the military action was perceived by some to be at least partly in response to the catastrophic failure of the international community to intervene in any effective way in the conflict in Bosnia. The NATO action provoked patriotism amongst even the most liberal anti-Milošević Serbs. The fall of Milošević followed mass demonstrations in Belgrade in 2000. Milošević was arrested and extradited to the ICTY in 2001 on charges of war crimes and genocide. He was found dead in his cell in The Hague before his trial was complete. Following the assassination of Serbian Prime Minister Zoran Djindjić in Belgrade in March 2003, the political scene in Serbia in the 2000s was dominated by perceptions of corruption, economic difficulties following years of sanctions and debate around the extradition of those indicted for war crimes. Former Bosnian Serb leader Radovan Karadžić who had evaded capture for 13 years was arrested in Belgrade and extradited to The ICTY.[8] Ratko Mladić, former Bosnian Serb military leader, was extradited in 2011.

Bosnia Herzegovina, since the conflict and at the time of writing, continues to consist of two entities, the (Muslim-Croat) Federation with its capital in Sarajevo and the (Serb) Republika Srpska with Banja Luka as its capital. The entities are governed by a tripartite system whereby each of the three constituent groups rotates the presidency. Each entity operates separate social and quasi-state structures, although some federal state structures are being slowly established. Minority returns of those displaced as a result of the conflicts continue to be generally slow with many areas having undergone massive ethnic shifts in population as a result of a deliberate policy of 'ethnic cleansing' on the part of Bosnian Serb forces. Kosovo declared independence in 2008 with Britain being one of the first to recognise the new state, amidst wide-spread protest by British Serbs. Slovenia was the first former Yugoslav republic to accede to the European Union in 2004 with Croatia joining in 2013.

Motivations behind the migration decision

When I embarked on the research and considered the different ways of approaching the analysis of the data I had collected, I had originally intended to explore whether an individual who had made the decision to migrate with a particular motivation exhibited qualitatively different transnational behaviour or affiliations from someone with another reason for migrating. The case study of the former Yugoslavia would offer, I believed, the opportunity to bridge the gap in the literature on the experiences of forced vs voluntary migrants by allowing for a transnational lens to be applied to the behaviours, practices and affiliations of those who had migrated for different reasons.

However, I quickly realised that conceptualising the myriad reasons for leaving the country of origin and coming to Britain was significantly more problematic than I had anticipated. In some cases, the migration experience was a straightforward decision made on the basis of a job offer or a university place for example, but often the questions 'why did you leave' and 'why did you come to Britain?' were followed by a sigh and then a deep breath before a narrative which contained within it a complex and multilayered labyrinth of emotion, pragmatism, choice and obligation. As the interviews progressed, I could see the neat spectrum of motivations I had envisaged disintegrating under articulations of resignation, fear, unhappiness, dissatisfaction, boredom, frustration, bewilderment, anger, resentment, bitterness and empowerment, pride, relief and gratitude.

As far as it is possible to classify, there were some broad 'categories' of motivation for migrating reflected in the respondents' narratives. There are for example the stories of those who came ostensibly to study, to further their career prospects or with existing offers of employment. Particular interest was shown in the project by a group of medical professionals from Serbia who, building on existing links established as far back as WWI between Serbian and British medical professions, had taken up clinical, research and teaching posts in Britain and who, through personal and professional contacts in Serbia, continue to facilitate a flow of medical and healthcare professionals from Serbia and the sharing of knowledge and skills training.

On a straightforward binary spectrum of migrating behaviour, those who came to Britain with firm career aspirations may be considered to be 'economic migrants'. And yet, the spectrum of motivations for migrating is rarely so unambiguous, especially in a part of the world where multiple, complex and protracted political and socio-economic factors can contribute to an individual migrant's decision to leave, can affect the migrant's relationship with the homeland and can ultimately inform the decision to return (or stay away). The degree to which it is possible to disaggregate any seemingly economic motivations from other circumstances in the region at the time of migration is debatable. Those who came with employment-related objectives could also be registering a dissatisfaction with the status quo in the local context, potentially unrelated to the job offer waiting for them in England. Mika, from Zagreb stated:

> I came for work basically. Well, that's kind of the reason I tell people. But actually, as a gay man, you don't get an easy time of it in the Balkans. I am much happier in London. I had a brilliant job back home but trying to be taken seriously was really difficult. My parents have always been ok with it. Sort of. But I have some seriously macho uncles. And to be

honest, I think it's better that am not around rubbing their noses in it. This way they can be proud of their successful son, nephew, whatever doing the fancy job in London and don't have to see me going out on a Saturday night with my friends.

(Croatian interviewee, male, age 27)

Bojana from Belgrade talked about her motivations for leaving:

I am a chemist and I work for the NHS. I get paid ok and it's enough to send back home to help out my parents a bit. They're retired now and they don't get much of a pension. I knew what I was coming for and I suppose if you had to say why I came, I am what the British government would consider an 'economic migrant'. Except am not. Not really. I just couldn't stay there anymore.

(Serbian interviewee, female, age 29)

Both Mika and Bojana are articulating how their apparent economic reasons for moving to Britain are underpinned by other motivations not solely related to economic rationales. For Mika and Bojana, the original catalyst behind their decision to migrate has, as time has passed, been superseded by the realisation that their lives would be qualitatively different in the homeland, and those differences would not only be financial in nature.

Day and White (2001: 18) draw on the work of Kunz (1973) who outlines two scenarios faced by the asylum claimant: the 'anticipatory' and the 'acute' scenarios, the main conceptual difference between the two being the speed and suddenness with which the departure occurs. In alluding to the greater intensity in international travel and migration in the latter part of the twenty-first century, Day and White add their conceptualisation of the 'blocked return' scenario where the international migrant who is studying or working outside of his/her country of origin is unable or unwilling to physically return as the situation in the country of origin deteriorates. There were a number of respondents who came to Britain with one (set of) motivation(s) but then found themselves essentially 'stuck' when the situation deteriorated in the region and full-blown conflict broke out.

I left the country before the war started to do research in Algeria where my late husband was also working on a temporary contract. When our contracts came to the end, the war in Bosnia was still going on and we could not return. During our visit to the UK in June 1994, we learned that all the international schools in Algeria will not re-open that autumn because other ex-pat families left and advised the schools that they do

not plan to return because of the terrorist threat in Algeria. My husband and I agreed that I should ask for political asylum in the UK.
(Survey respondent, female, identifies as Yugoslav,
left country of origin at the age of 40)

The story of Lara and her family above illustrates how Day and White's 'blocked return' scenario can be further complicated when the experience of being 'blocked' relates to both the country of origin and the first country of settlement.

Azra from Mostar talked about how she stayed in London, despite her original plans to return. Her experience echoes the stories of others who articulated how their return had been advised against by those left behind, those who were coping with the daily realities of life at 'home':

I came as an au-pair but then when it all kicked-off over there I didn't have much choice but to stay. I didn't want to actually, I was all up for going back but my parents said absolutely no way, they ended up sending my sister here as well then too. It was hell by the sounds of it and they just didn't want me going back to that.
(Bosnian interviewee, female, age 27)

In the cases of many of those from Bosnia and Kosovo there was little ambiguity over the motivation for leaving the country of origin even if the question of whether/when to return was more nuanced. The following testimonies are from those who came to Britain for no reason other than the ongoing violence or threat of threat of violence in their homes:

My City Sarajevo was under complete siege by Serbian forces, therefore it was impossible to leave the City at the time. I left Bosnia by UN flight with the help from the TV agency I was occasionally working for at the time. It was a scary experience as I had to go through the Serbian forces check-point in a UN forces protected vehicle.
(Survey respondent, male, Bosnian, arrived in
Britain in 1993 at the age of 28)

Left my country because of the Serbian repression and the fear that my life was in danger.
(Survey respondent, male, Kosovar, arrived in
Britain in 1997 at the age of 22)

I was not a Muslim from Bosnia and I had great problems leaving ex-Yugoslavia countries. Although I was raped I couldn't provide details because no-one was bothered about vulnerable women who were supposed to be Orthodox. In Serbia I couldn't register as a refugee because my father was Macedonian, in Macedonia I couldn't first enter because I had Bosnian ID card, and then I couldn't get any help because they understood me that I am Macedonian and not a refugee.

> (Survey respondent, female, identifying as mixed ethnicity, from Yugoslavia, left Sarajevo at the age of 22 in 1992, arrived in Britain in 1999)

I left Bosnia before the war to study in Slovenia. When the war started in Bosnia, our home was destroyed and my father was taken into concentration camp 'Manjaca', my mother and younger sister were forced to leave Sanski Most. They found refuge in Velika Kladusa. In November 1992 my father was released from the concentration camp under the protection of UNHCR and International Red Cross to Croatia, then from Croatia in December 1992. His flight was organised to the UK under government programme '1000'. I left Slovenia in early February 1993 to Croatia where I joined the group for flight to the UK as part of family reunion programme. In March 1993 my mother and younger sister arrived to the UK and we reunited finally after 2 years.

> (Survey respondent, female, Bosnian, arrived in Britain in 1993 at the age of 18)

One interviewee, Hana, explained how finding a job was becoming more difficult for her as the situation changed in Serbia and how her career was essentially being affected by her 'otherness':

I was having some problems basically because of my background. I was teaching at university in Novi Sad and I kept being passed over for promotion as not being '100%' Serb. So I thought it would be good to switch to Belgrade. There was an opening at the university which I thought would be good for me but then during the questioning my surname was recognised as being Croat. Serbian TV was awful at this time, I did some research on the effects of the wars on children. I had noticed that even in pre-school children they were mirroring what they saw on TV, drawing pictures of blood, the ways they played were changing, a lot of war games and that kind of thing. Parents just weren't able to help their children understand what was happening. Anyway, I presented all of this and called for more counselling services and the reaction was not good. The audience said my research wasn't objective enough. And then at the same conference a colleague of mine quoted a paper written by a Slovenian psychologist. And someone stood up and complained about 'foreign' sources being used at the conference. I knew then that was the end of my career there, there was no way I could work under those conditions.

(Serbian interviewee, female, age 67)

Not all of the participants in the research came during the 1990s however. The narratives of those who made their journeys earlier are interesting to explore for their myriad motivations for coming to Britain initially and then deciding to stay and make their lives here. Ana from Serbia has the following story:

I was always very good at languages even when I was very young and then when I finished university I got a job working in the Ministry for Foreign Trade. I was the only breadwinner in my family and my father was very ill at that time. So, when I had finished my studies, they called me one day – the Ministry and they said to me 'we want to send you to London to be an interpreter for us'. 'You are our future' they said to me. So, I came in 1961 to do a course. There was one Saturday when my two friends made me go out to a lovely dancing venue. So, we were there, and he, my husband to be, asked me to dance. And then he called me on the phone one day and he asked if I would meet him. And then just before I left he gave a party for me and my friends. When the time came and I had to go back to Belgrade by train I thought I would never see him again. I was a member of the Communist party after all and they had paid all that money for me. Then, I couldn't believe it. I was at home with my parents back in Belgrade and there was a knock on the door and there he was! My father was a dying man by this time. And I have never forgotten what he said, he said 'he looks like a good man'. It was against the law at that time to marry a foreigner. He stayed with us for one week and then he left. Our local registrar said I would be jailed if I married him. So, what to do? Well, Tito's cousin was a very good friend of our family. I told him that I wanted to go to Trieste to buy some clothes. He took me to the Ministry to get a visa. I waited until I was alone in the office and then I told them that I wanted to go abroad to London. The time then waiting to go passed so slowly. I remember waving goodbye to my mother and my sister and I thought that was it, I thought I would never see them again. There was a very famous Serbian classical actress on the train with me in my compartment, we were talking and she told me that I was stupid and that I'd burnt all my bridges. He was waiting for me at Victoria station when I arrived – biting his fingernails. Later on I found out that when the train was in Austria someone had told them in my office where I was really going and they tried to chase after me but the train had already crossed the border.

(Serbian interviewee, female, age 68)

Another Croatian interviewee talked about how she had essentially made the move as a younger woman to London after meeting her (British) husband-to-be in Zagreb.

I was very reluctant to come here to live. I had an interesting job I loved working for a museum in Zagreb, then I met my (British) husband at an exhibition and we moved to Camden. I knew it would be difficult for me with a young child and not knowing anyone.

(Croatian interviewee, female, age 56)

Branko from Serbia who came to England after WWII as a child gave his family's story:

My mum was the daughter of a greengrocer and then she got pregnant with me. There was no chance in those days of not doing the 'honoura-ble' thing, so they got married. Then my Dad who was with Mihailović in the war ended up in a camp in Italy. To be honest, I think it was a bit of 'out of sight, out of mind' with my dad. I don't honestly reckon he was that bothered if my mum knew whether he was dead or alive. He never bothered to get in touch or anything to let her know he was alright. It was kind of funny actually how she found out. He'd been boasting to some guy in the camp that he had married the daughter of this famous greengrocer and then what happens? The guy's wife is practically our next-door neighbour. So, she tells my mum and the next thing we knew we were off to London. We just ended up there because that's where he had ended up. There's this story that the first Saturday we were there, my dad was sitting drinking with all his mates as usual at the Serbian club, my mum was doing the dishes with all the other women and she went and tapped him on the shoulder and he looked totally confused, like 'who the hell are you?' He had basically forgotten all about us. With him I reckon he was happier with the photos in his pocket than he was with the reality of his family. It happened to lots of families like that during that war. It happened to a tenant of ours, back in Belgrade. He was only 17 and one day he got pulled off his horse and ended up in a camp somewhere, they were reunited years later, but not really if you know what I mean. By the time his family found out he was alive they had kind of forgotten all about him. He died in the end as a pensioner in a one-bedroom flat in London and it was a week before anyone discovered his body.

(Serbian interviewee, male, age 62)

And then there were the articulated experiences of those who made the move to Britain not fleeing war or violence, or attempting to build a career:

I was running away from my parents, that was my main motivation to be honest, anywhere as far away from them as possible would do.

(Serbian interviewee, female, age 33)

The extracts above have demonstrated the diversity of catalysts behind migration from the former Yugoslavia to Britain. Participants in this project made the decision to leave their countries of origin for reasons of violence or the threat of violence, discrimination on the grounds of ethnicity or sexuality, to pursue study or career opportunities as a means of demonstrating a rejection of the dominant political discourse, for love and marriage and to assert their independence from family members. The reasons why they chose to remain and not to return are equally complex, ever-changing and above all difficult to encapsulate with one 'label' or theory. What they all had in common, however, by virtue of their status as a 'migrant', be it 'forced', 'voluntary' or anywhere on a non-binary spectrum in-between, was one distinct moment of departure from the country of origin. For some, those departures are experienced in repeat on subsequent visits to the homeland but reflected in so many of the narratives were clear articulations of myriad emotions related to the 'original' departure, sometimes reoccurring in dreams years later.

Reflecting on the experience of departure(s)

Miroslav Jančić (1996), in his collection of poetry, wrote 'Some people say, Life consists of departures'. The points of departure from the homeland into (often) an indefinite period of exile acted as punctuation to the narratives of some respondents in a tangibly painful way. The 'first' departure when the initial act of migration took place was usually the catalyst for expressed emotions of overwhelming sadness, loss, panic, regret, guilt but in the narrative of some migrants we can see these emotions being played out in a no less intense way at the telling of subsequent less obviously 'significant' departures. This was a theme common across all groups, regardless of any demographic differences and time of arrival in Britain.[9]

I can't talk about it actually. It is something that absolutely haunts me, that moment. I have been back plenty of times, there have been lots of departures but that one, when I didn't know if I was going to see any of them again [...] And I knew then that was it for us, our lives were wrecked even if by some chance we were all going to be ok.

Many of those interviewed, when asked when they arrived in Britain, responded with the exact date of arrival, even those who arrived decades ago. Some described their moment of departure in detail and others preferred not to talk about it. One interviewee talked about how she lives for her visits 'home' but describes how every time she has to leave to return to England, she re-enacts the feelings of the 'original' departure.

Of course nothing is as bad as that original time, the chaos, the tears but it's always there, every time I go to the airport, I remember how it was. So, so painful. The night before I have to come back to England – every time – I get this horrible feeling in the bottom of my stomach and I don't sleep.

We are able to see in these narratives how the physical act of departure and the memory of other departures are particularly intensified around visits to the homeland, which some deal with by avoiding altogether. This further demonstrates how counting the frequency of visits home, so often considered to be a key feature of transnational activity, should not be considered to automatically be evidence of transnational character or commitment.

Associated with departures and other motifs running through the narratives of migration is an underlying current of guilt, which was overwhelmingly heard in the different voices articulating their migration stories.[10] We have seen how the element of 'choice' is present in some narratives more than others, nuances around which can call into question the distinction between who is to be determined as a 'forced' or 'voluntary' migrant. The following chapter discusses some of the ways in which such nuances can become challenged within the political and legislative context around migration in Britain – into which those from the former Yugoslavia would have been arriving at different times.

Notes

1 See Chapter 3.
2 Royalist Serb general during WWII and Chetnik leader.
3 For example, the Ravna Gora society (and hotel).
4 During the late 1980s shortages of staple goods were becoming more common. The early 1990s saw a period of hyper-inflation in Bosnia and in Yugoslavia savaging what was already a struggling economy. Rates were high enough at one

stage to invoke reports of wages being paid in cash to workers, and the wages being worthless by the time the recipient could get to a shop to spend them. In 1994 the rate hit its highest of a daily inflation rate of 64.63 per cent, taking 1.4 days for prices to double. The infamous symbol of this accumulation was the production in 1993 of a 500 billion Yugoslav dinar banknote (an example of which can be viewed at the British Museum).

5 'Bosnian researcher counts war dead and faces threats for this methods', Radio Free Europe, Radio Liberty, 21 November 2008.

6 See Baker (2015) for an overview of the ways in which different analysts have approached the conflicts, including a comprehensive reading list; Gordy (2008: 28) for his identification of 'seven moments [...] marking the break at which the continued existence of Yugoslavia as a federal state [...] was impossible or extremely improbable'; and Dragović-Soso (2008) for her analysis of five groups of explanations within the literature on the disintegration of Yugoslavia.

7 See Balicki and Wells (2006: 7) for details of the survey of Kosovar families in east London.

8 In March 2015, Karadžić was convicted by the ICTY of genocide, war crimes and crimes against humanity and sentenced to 40 years of imprisonment.

9 I have therefore not identified respondents here by any demographic factor, such as ethnicity, gender, age.

10 I discuss this further in Chapter 3.

References

Aldridge, O. (1916) *The Retreat from Serbia through Montenegro and Albania.* London: Minerva.

Baker, C. (2015) *The Yugoslav Wars of the 1990s.* London: Palgrave Macmillan.

Baletić, Z. (1982) 'International migration in modern economic development: With special reference to Yugoslavia', *International Migration Review* 16(4): 736–756.

Balicki, J. and Wells, A. (2005) 'Asylum seekers' family wellbeing as a challenge for immigration policies'. Paper delivered at the International Union for the Scientific Study of Population Conference, Tours, France, July 18–23, 2005.

Balicki, J. and Wells, A. (2006) *Asylum Seekers' Policy vs. Integration Policy.* London: Trafford Publishing.

Baučić, I. (1974) 'Yugoslavia as a country of emigration', *CHIEAM, Options Méditerranéennes* 22: 55–66.

Behloul, S. (2011) 'Religion or culture? The public relations and self-presentation strategies of Bosnian Muslims in Switzerland compared with other Muslims', in M. Valenta and S. Ramet (eds) *The Bosnian Diaspora: Integration in Transnational Communities.* Surrey: Ashgate Publishing.

Berry, J. and Berry, F.M.D. (1916) *The Story of a Red Cross Unit in Serbia.* London: J&A Churchill.

Bieber, F. (2002) 'Nationalist mobilization and stories of Serb suffering', *Rethinking History* 6(1): 95–110.

Carter, S. (2007) 'Mobilising generosity, framing geopolitics: Narrating crisis in the homeland through diasporic media', *Geoforum* 38: 1102–1112.

Čapo-Žmegač, J. (2007) *Strangers Either Way: The Lives of Croatian Refugees in their New Home* (translated by Nina H. Antoljak and Mateusz M. Stanojević). New York: Berghahn.

Čolić-Peisker, V. (2008) *Migration, Class and Transnational Identities: Croatians in Australia and America*. Urbana: Illinois University Press.

Coughlan, R. and Owens-Manley, J. (2006) *Bosnian Refugees in America: New Communities, New Cultures*. New York: Springer.

Dahinden, J. (2005) 'Contesting transnationalism? Lessons from the study of Albanian networks from former Yugoslavia', *Global Networks* 5(2): 191–208.

Dahinden, J. (2009) 'Understanding (post)-Yugoslav migration through the lenses of current concepts in migration research: Migrant networks and transnationalism', in U. Brunnbauer (ed.) (2009) *Transnational Societies, Transterritorial Politics: Migrations in the (Post) Yugoslav Region, 19th–21st Century*. Munich: R. Oldenburg Verlag GmbH München.

Day, K. and White, P. (2001) 'Choice or circumstance: The UK as the location of asylum applications by Bosnian and Somali refugees', *GeoJournal* 55: 15–26.

Dragović-Soso, J. (2008) 'Why did Yugoslavia disintegrate? An overview of contending explanations', in L.J. Cohen and J. Dragović-Soso (eds) *State Collapse in South-Eastern Europe: New Perspectives on Yugoslavia's Disintegration*. West Lafayette: Purdue University Press.

Eastmond, M. (1998) 'Nationalist discourses and the construction of difference: Bosnian Muslim refugees in Sweden', *Journal of Refugee Studies* 11(2): 161–181.

Fazlić, H. (2009) 'First Bosnians in the UK'. Birmingham: Bosnian Cultural Centre.

Foster, S. (2013) 'British medical volunteers and the Balkan Front 1914–1918: The case of Dr Katherine Stuart MacPhail', *University of Sussex Journal of Contemporary History* 14: 4–16.

Fozdar, F. (2009) '"The Golden Country": Ex-Yugoslav and African refugee experiences of settlement and "Depression"', *Journal of Ethnic and Migration Studies* 35(8): 1335–1352.

Franz, B. (2002) 'Bosnian refugees and socio-economic realities: Changes in refugee and settlement policies in Austria and the United States', *Journal of Ethnic and Migration Studies* 29(1): 5–25.

Franz, B. (2005) *Uprooted and Unwanted: Bosnian Refugees in Austria and the United States*. College Station: Texas A&M University Press.

Gordy, E. (2008) 'Destruction of the Yugoslav Federation: Policy or confluence of tactics?', in L.J. Cohen and J. Dragović-Soso (eds) *State Collapse in South-Eastern Europe: New Perspectives on Yugoslavia's Disintegration*. West Lafayette: Purdue University Press.

Halilovich, H. (2011a) 'Beyond the sadness: Memories and homecomings among survivors of "ethnic cleansing" in a Bosnian village', *Memory Studies* 4(1): 1–11.

Halilovich, H. (2011b) '(Per)forming 'Trans-local' homes: Bosnian diaspora in Australia', in M. Valenta and S. Ramet (eds) *The Bosnian Diaspora: Integration in Transnational Communities*. Surrey: Ashgate Publishing.

Halilovich, H. (2013) *Places of Pain: Forced Displacement, Popular Memory and Trans-Local Identities in Bosnian War-Torn Communities*. New York: Berghahn.

Halilović-Pastuović, M. (2007) 'The "Bosnian project": A "vision of divisions"', in B. Flanning (ed.) *Immigration and Social Change in the Republic of Ireland.* Manchester: Manchester University Press.

Halpern, J. (1975) 'Some perspectives on Balkan migration patterns (with particular reference to Yugoslavia)', in B.M Du Toit and H.I. Safa (eds) *Migration and Urbanization: Models and Adaptive Strategies.* The Hague: Mouton Publishers.

Herlihy, J., Scragg, P. and Turner, S. (2002) 'Discrepancies in autobiographical memories – implications for the assessment of asylum seekers: Repeated interviews study', *British Medical Journal* 324: 324–327.

Jančić, M. (1996) *The Flying Bosnian: Poems from Limbo.* London: Hearing Eye.

Jansen, S. (2008) 'Misplaced masculinities: Status loss and the location of gendered subjectivities amongst 'non-transnational' Bosnian refugees', *Anthropological Theory* 8(2): 181–200.

Judah, T. (1997) *The Serbs: History, Myth and the Destruction of Yugoslavia.* Yale: Yale University Press.

Kelly, L. (2003) 'Bosnian refugees in Britain: Questioning community', *Sociology* 37(1): 35–49.

Kelly, L. (2004) 'A community empowered? The Bosnia project in the UK' in P. Essed, G. Frerks and J. Schrijvers (eds) *Refugees and the Transformation of Societies.* Oxford: Berghahn.

Kosińksi, L.A. (1978) 'Yugoslavia and international migration', *Canadian Slavonic Papers/Revue Canadienne des Slavistes* 20(3): 314–388.

Kostovicova, D. and Prestreshi, A. (2003) 'Education, gender and religion: Identity transformations among Kosovo Albanians in London', *Journal of Ethnic and Migration Studies* 29(6): 1079–1096.

Kunz, E.F. (1973) 'The refugee in flight: Kinetic models and forms of displacement', *International Migration Review* 38: 91–11.

Lampe, J. (1996) *Yugoslavia as History: Twice There Was a Country.* Cambridge: Cambridge University Press.

Leneman, L. (1994) *In the Service of Life: The story of Elsie Inglis and the Scottish Women's Hospitals.* Edinburgh: The Mercat Press.

Lerch, M., Dahinden, J. and Wanner, P. (2006) 'Remittance behaviour of Serbian migrants living in Switzerland: A survey'. Swiss Forum for Migration and Population Studies.

Liddington, J. (2011) 'Britain in the Balkans: The response of the Scottish Women's Hospital Units', in I. Sharp and M. Stibbe (eds) *Aftermaths of War: Women's Movements and Female Activists 1918–1923.* History of Warfare, Volume 63. Boston: Brill Leiden.

Mikić, Ž. (2007) *Ever Yours Sincerely: The Life and Work of Dr. Katherine S. MacPhail* (translated by Dr Muriel Heppell). Cambridge: Perfect Publishers.

Miller, L. (2012) *A Fine Brother: The Life of Captain Flora Sandes.* London: Alma Books.

Mišković, M. (2011) 'Of home(s) and (be)longing: Bosnians in the United States', in M. Valenta and S. Ramet (eds) *The Bosnian Diaspora: Integration in Transnational Communities.* Surrey: Ashgate Publishing.

Povrzanović Frykman, M. (2011) 'Connecting three homelands: Transnational practices of Bosnian Croats living in Sweden', in M. Valenta and S. Ramet (eds) *The Bosnian Diaspora: Integration in Transnational Communities*. Surrey: Ashgate Publishing.

Procter, N. (2000) *Serbian Australians in the Shadow of the Balkan War*. Hants: Ashgate Publishing.

Pryke, S. (2003) 'British Serbs and long distance nationalism', *Ethnic and Racial Studies* 26(1): 152–172.

Ramet, S. (1996) *Balkan Babel: The Disintegration of Yugoslavia from the Death of Tito to Ethnic War*. Boulder, CO: Westview Press.

Silber, L. and Little, A. (1996) *Yugoslavia: Death of a Nation.* London: Penguin.

Skrbiš, Z. (1999) *Long-Distance Nationalism: Diasporas, Homelands and Identities*. Hants: Ashgate.

Slavnić, Z. (2011) 'Conflicts and inter-ethnic solidarity: Bosnian refugees in Malmö', in M. Valenta and S. Ramet (eds) *The Bosnian Diaspora: Integration in Transnational Communities*. Surrey: Ashgate Publishing.

Summerfield, D. (2003) 'War, exile, moral knowledge and the limits of psychiatric understanding: A clinical case study of a Bosnian refugee in London', *International Journal of Social Psychiatry* 49: 264–268.

Turner, S. (2003) 'Mental health of Kosovan Albanian refugees in the UK', *The British Journal of Psychiatry* 182: 444–448.

Valenta, M. and Ramet, S. 'Bosnian migrants: An introduction', in M. Valenta and S. Ramet (eds) *The Bosnian Diaspora: Integration in Transnational Communities*. Surrey: Ashgate Publishing.

Wight, E. (2000) 'Bosnians in Chicago: Transnational activities and obstacles to transnationalism'. Sussex: Sussex Centre for Migration Research.

2 Contexts of arrival and reception

The concept of the 'context of reception' as defined by Portes and Rumbaut (2006) and further developed by Stepick and Stepick (2009) and Jaworsky *et al.* (2012) is one which has been applied to a number of empirical studies but less frequently within a European context and often with a focus principally on the labour market and economic variables. The principles of the context of reception are predicated on the assumption that certain aspects of the host country's socio-political environment will influence a migrant's 'incorporation' and identity formation as a member of the host society. Portes and Rumbaut (2006) identify those socio-political aspects as: policies of the host government, conditions of the host labour market, characteristics of the new arrivals' co-ethnic communities in the host country and the reactions of other communities of different ethnicity from the migrant. Other scholars have referred to similar variables without conceptualising them under the umbrella of the 'context of reception'. Kelly (2003) for instance, in her study of Bosnian refugee 'communities' in Britain explored the different ways in which migrants from Bosnia were treated by the host society, depending on whether they arrived independently or whether they arrived as part of a sponsored (for example UNHCR) programme. Again within the US context, other research has emphasised the role of media and their representation of migrants (Padin, 2005) and how local political action (Bloemraad, 2004) can have an influence on the position in which the new migrant will find him/herself.

What much of the literature has in common is the focus of the context of reception in relation to 'integration' or 'incorporation' of the migrant community into the host society. I have deliberately avoided the term 'integration' and its frequently accompanied term in the literature 'assimilation' as concepts which I find difficult to apply to the pragmatic realities of the lives of people as they reported their experiences during the course of the research. In a society and environment which is as diverse as those in which some migrants find themselves in Britain, particularly in large cities, it would be difficult to define what is meant by integration. Observers have highlighted for instance how London would be a very different context of reception in which migrants would find themselves compared with other parts of Britain (Bajić-Hajduković, 2008; Kershen, 1997).

The host environment is not static and where newcomers find themselves positioned when they first arrive will change over time as both they and their society evolves. Indicators of integration as defined by governmental policies often do not lend themselves to a more nuanced understanding of how someone born in a different country from which he/she is now resident positions him/herself in terms of individual identity, community identity and relationship with his/her country of origin. A focus on the context of reception solely in the host society with an emphasis on the apparent goal of 'incorporation' risks missing a crucial aspect of the migration experience. A migrant is not arriving out of a vacuum into the host society; he/she brings a varying degree of life experience and in some cases, will have already experienced the majority of his/her life cycle within a different environment.

Arrivals in Britain

Quantitative data on migration from the region to Britain is difficult to come by. Much of the governmental statistical evidence refers to data on arrivals from Yugoslavia and is difficult to disaggregate further into data on the different constituent republics. Despite efforts by the British government to implement dispersal policies, in common with other migrant groups, many of those from the former Yugoslavia have settled in the London area.

Census data illustrates the impact which the conflicts in the region had on migration to Britain. According to the census, residents in Britain (England, Scotland and Wales) reporting their country of birth as one of the former Yugoslav countries increased from 13,846 census respondents in 1991 to 47,410 in 2001. A study on the impact of conflicts overseas on UK communities (Collyer *et al.*, 2011) quoted 'ethno-community estimates', which it indicated were 'likely to be an under-estimate of actual population size'. The figures cited in that report were 'Bosniacs'[1] – 8,000–10,000; Kosovar Albanians – 50,000; Serb[ian]s – 70,000; Croats – 7,500. Similar figures were cited in my interviews with community representatives.[2] Slovenian national law dictates that its diaspora members have to register with the Slovenian Embassy which explains why the Slovenian Embassy representative could state with more confidence that at the time of the interview they had a register of 530 'permanent' Slovenian residents in Britain with approximately 3,000 on their email database, many of these being students. Both Slovenians and Croatians have vibrant student societies, especially in London.

There are a number of possible explanations for the discrepancy between the census and those figures quotes by community representatives. Community group interviewees were responding to the way the question was framed during interviews. Respondents were, in the main, quoting estimates they have for ethnicity, rather than nationality. The community representative estimate for Serbs for example would be referring to Serbs of all nations, not just the state of Serbia and would also include Serbs who were born in Britain but identify as Serb by ethnicity. The census on the other hand is asking a much narrower question around country of birth.

Interviews with community representatives indicate that in general: most Bosnians and Kosovars in Britain today would have migrated as a result of the conflicts in the 1990s; many Slovenian migrants are here for educational or career reasons and are likely to return to Slovenia; and Croatian migrants represent a mixed group of students, professionals and those fleeing the conflict in Croatia. Bosnian and Slovenian community representatives also refer to 'small numbers' of migrants from Bosnia and Slovenia coming to Britain during the post-WWII period. Of the ethnicities or nations which made up what was Yugoslavia, Serbs or Serbians have experienced the longest history of migration to Britain and would, therefore, be expected to have had the opportunity to have developed longer-term networks and/ or communities. There are several thriving Serb community groups, with different aims: for example, the Serbian Society, a registered charity with broadly speaking social and cultural activities; the Serbian Council of Great Britain which is more politically motivated; and the Serbian City Club – a network of Serb professionals working in financial institutions within the City of London.

In tracking the history of migration from Serbia to Britain, Ninković has distinguished between the temporary migration of pro-European educated elites between 1860–1940, coming to Britain to capitalise on education and training opportunities before returning to Serbia and Yugoslavia with newly-acquired knowledge and experience in the areas of philosophy, economics and technology. The Serbian Relief Fund also arranged for the education in Britain of 'some hundreds of Serbian boys' during WWI.[3] The second post-1940 period marked by major political catalysts – WWII, the rise of the Communist regime, the violent dissolution of Yugoslavia, economic collapse and sanctions – imposed a more permanent state of exile on the majority of migrant Serbs and indeed other migrants from the region. Mavra (2010) identifies three main waves of migration of Serbs to Britain – the first being made up of exiles from the new Communist regime in Yugoslavia after WWII and European Voluntary Workers (EVWs),[4] the second Mavra terms 'economic migrants' leaving Yugoslavia between the 1960s and late 1980s and then refugees fleeing the conflicts of the 1990s merged with more recent highly skilled workers, students or those seeking better economic and employment opportunities. Comments made during an interview with one Serbian community representative indicated that the 1980s wave of Serbian migrants 'wanted little to do with those who came earlier – labelled as "defectors" and "collaborators"'.[5]

Although framing the migration experience in terms of waves or patterns of migration by date of departure from the country of origin can sometimes be useful to contextualise the debate, the diversity of the various catalysts behind migration decisions on the part of individuals and families from different parts of the former Yugoslavia is so intense as to render a chronological framing one-dimensional. In a region which has experienced such a concentration of (often) localised trauma in a relatively short period of time, a temporal analysis of migration motivations and experiences by ethnicity is not always helpful in itself, without a simultaneous consideration of other factors.

Negotiating the British immigration system

The legislative and political framework around immigration and asylum in Britain is a fast-moving environment and is one which I do not attempt to cover extensively here. In a similar way to the discussion of the historical context of the former Yugoslavia discussed in Chapter 1, I have couched this presentation of the immigration discourse in Britain within my respondents' experiences of the formal migration processes and procedures.

The majority of my respondents arrived in Britain between 1990–1999, independently and did not migrate under the sponsorship or guidance of any organised programme. Prior to the conflicts in the 1990s, migrants from the region made the journey on (primarily) work or student visas. Following the break-out of fighting and especially in reaction to the horrific images shown via media reports of (mainly) Bosnian Muslims being held in concentration camps, Britain's response under international 'burden-sharing' agreements took the form of the 'Bosnia Project' – a means of administering support for a quota of 1,000 refugees who made the journey under the auspices of the UNHCR and the Red Cross, many of whom were former detainees of the camps, some later to be joined by family members. The temporary nature of the protection offered to refugees from Bosnia, and later from Kosovo, was not just a British policy response but was also a solution offered in other European and non-European states (King, 2001). 'Look and see' programmes were initiated whereby refugees were encouraged to return for a set period to their home country as, presumably, a type of reassurance measure and to encourage more permanent return (Walsh *et al.*, 1999). Kosovan refugees to the UK were kept up to date on events 'back home' through the magazine *Lajmetari*, a publication issued by the Refugee Council (Smart, 2004: 26).[6] In the case of Bosnian and Kosovar migrants, some, but by no means all, of the 'programme' or 'quota' refugees returned to their homes in the region.[7]

The impact of temporary protection policies offered to forced migrants upon the refugee experience in the host country and the associated feelings of disempowerment have been discussed within the European (Koser & Black, 1999; Kjaerum, 1994) and wider context (Mares, 2001). From the sample of migrants from the former Yugoslavia who engaged with my research I found that a clear demarcation between those migrants who arrived 'voluntarily' and those who were 'forced' was not always possible. The reported experience of some was relatively straightforward with no significant problems, however for many, the wait for a decision over their immigration status dominated every aspect of their lives. This unfortunate experience of the British immigration system was a feature of many migrating in the 1990s, regardless of region of origin or motivation behind the decision to migrate. The length of time that it took the immigration authorities to communicate a decision on the applicant's immigration status was crucial to the possibilities of gaining access to education, training, the ability to take paid legitimate employment and to the general opportunity to make decisions, choices and have control over a life and a future. Amongst my research participants, for those who did not come with a work or study-related visa, there was an average of a seven year wait from the Home Office for a decision on applications for immigration status. The longest length of time reported spent waiting was 13 years. For many this would have meant a corresponding temporary renunciation of their passport or travel document to the Home Office for all or at least part of the time waiting, making travel abroad a physical impossibility.

One Serbian young woman, who had come to London as an au pair in 1992, experienced years of limbo waiting for the immigration authorities to issue her with a travel document. Her younger brother was 12 years old when she left Serbia at the age of 19 and he was 22 by the time she was able to see him again. Another respondent, a Bosnian Croat who came to London, again on an au pair visa in 1991, was advised by her parents not to return after conflict broke out in Croatia and then in Bosnia. Her youngest sister was two years old when she left and was aged nine when she was able to see her next. Contact via social media at that time was not as ubiquitous as it is today and regular contact even by telephone would have been difficult at times because of the problems in telecommunication networks. Enforced fractures in the familial bonds meant for some that those relationships were never fully recovered or in some cases had not even had a chance to have fully developed. Indeed, several respondents expressed how they felt as though other members of the diaspora had taken on the role of 'family substitute', in terms of the development of a multi-layered familial dynamic.

For those seeking asylum and for those who came to Britain with a certain immigration status which changed as the situation worsened in the region, this enforced period of 'limbo' was one which, despite efforts to fight against it, has formed a backdrop of frustration which has tinged much of their subsequent perceptions of British authorities and has crystallised into a tangible sense of loss at, in some cases, many wasted years of opportunity.

> It really is one of the most difficult countries to emigrate to. You are supposed to get an education and then that sets you up in your career. But I had no chance of that, it's been a wasted life – I didn't enjoy anything, I didn't travel, I have always been playing catch-up.
>
> (Croatian interviewee, female, age 40)

Echoes of a similar attitude towards the immigration authorities were found in the following extract:

> The Home Office became a term of disgust and aversion amongst my friends. […] Until I got my visa, I would shiver every time there was a mention of the Home Office. There was terror, this place looks like haunted and basically the problem is solely bureaucracy and nothing else, people who deal with the problems. Prisoner in the sense that I have all the rights, I have the right to live, I have the right to buy and spend money in the country, I have the right to study but I do not have the right to travel unless they say so.[8]

There were many stories of incompetent immigration solicitors told by respondents, regardless of country of origin, ethnicity or status applied for, combined with sometimes an expressed level of disgust for the Home Office so intense that that institution almost takes on an identity of its own amongst the narratives of some respondents. Serbian interviewee, Mira stated:

> The very thought of the Home Office, even now, it is amazing how two words can cut into one's blood, nature to the extent that it becomes so bad that the mere thought of it produces some kind of sickness, it is simply unbelievable.
>
> (Serbian interviewee, female, age 38)

It is possible to see clearly through respondent narratives how the implications of temporary protection have reinforced the rupture of the forced departure in terms of life trajectories and, in some cases, have had unequivocal and overwhelming effects on the ways in which individual migrants had to adapt any future life plans at both the level of the individual and the family. Such periods in limbo also had more insidious effects in terms of discouraging any sense of entitlement, control or confidence in relation to the homeland and particularly concerning the new home which, as is made clear by the immigration authorities, may only be a temporary refuge – a refuge not worthy of emotional investment and offering limited opportunity for practical development. Through the articulation of such experiences the transnational space is not a comfortable place to be a migrant.

Vathi (2010), in her study of Albanian and Kosovar Albanian migrants in London emphasised the effect that a long period of waiting for a decision on immigration status has had on the mental health of those she spoke with, conceptualised by Kostovicova and Prestreshi (2003) as 'uncertain diasporans'. Bosnian poet, Miroslav Jančić, offers the following thoughts about the effects of undecided status, linked to the protracted fighting in Bosnia and unknown outcomes of the war, on those fleeing the conflict.

Kind of collapse
It's this idleness
Which is killing me most
An eternity of temporariness
The awareness of the billions of years
Of work and its worthy achievements
Transformed into the ashes
Of that inane war.
                    (Jančić, 1996: 34)

The temporariness and feelings of limbo as referred to by Jančić combined for some with the difficult circumstances in which they were forced to leave their homes to form a sense of paralysis, reinforced by the immigration authority's approach.

An almost total absence of choice and control over the circumstances in which they were forced to flee the homeland led in many cases to a similar lack of agency in the months and years immediately following arrival in Britain. Some narratives show how it was not only problems with the immigration authorities which could result in feelings of impermanency and uncertainty about the future. One Croatian interviewee coming to Britain in the 1990s and who had had the good fortune to have an uncomplicated experience within the immigration system discussed how she had also felt restricted over the lack of clarity about her future, a confusion which may not have been imposed by any external authority but was still having an effect over her life:

> After seven years I mentally decided that my home was going to be the UK. My life was basically static until then, it was only once I had made a decision that I could get on with things and stop being in limbo.
>
> (Croatian interviewee, female, age 38)

One interviewee highlights how the perceived temporary nature of the state of migration was an issue also for those migrants coming at earlier times. Whilst the policies of recent British governments towards asylum have had an impact upon the longer-term settlement decision-making processes of those fleeing the conflicts of the 1990s, those migrating within a different set of circumstances may also have felt at the mercy of wider political forces outside of their control. Serbian migrant Branko who came to Britain as a child in the wake of WWII explained the feelings of his parents and those in a similar position who had come to Britain as royalist followers of chetnik leader Mihailović:

> None of us had a clue that we would be here for that long. My father always promised my mother that we would return as soon as he [Tito] fell. It was the same with all of the families in the same position as us. They never intended to make their lives here, nobody thought he would go on as long as he did.
>
> (Serbian interviewee, male, age 62)

Observers have highlighted how the acquisition of secure immigration status can have an impact upon the development of transnational ties (Mazzucato, 2007). At the level of both the individual and collective, secure status within the 'new environment' can help to develop a degree of confidence and assurance which can then have an impact upon the relationship that the individual migrant, and any community which they are a part of, may develop with the homeland and the new home. In this way the transnational space can be both directly and more obliquely shaped and influenced by the migrant perception of status and security. Such status and security could be interpreted at multiple different levels, including the ways in which the country of origin and its communities are represented within the public discourse.

Representations of the former Yugoslavia and its migrating populations

As much as security of immigration status affects the transnational actions and character of an individual migrant, 'security' is not restricted to a decision made by the immigration authorities. Financial, employment, housing, education, relationship factors will all have an impact upon the 'security make-up' of any individual – migrant or non-migrant. One of the factors which contributes to sufficient degrees of collective 'self-confidence' leading to mobilisation significant enough to challenge the status quo is the ways in which some communities are viewed through the eyes of members of the elite within the new homeland. In this regard the Serb collective in Britain is significantly more 'advanced' than others from the former Yugoslavia, partly because of the history of migration from the region and partly because of certain relationships evidenced through state connections and influence. Any evidence to suggest that those relationships may be in jeopardy have led to concentrated efforts on the part of the Serbian collective in Britain to safeguard that level of assurance. Within such a dynamic, the transnational space can become an active and politicised environment. I detail here some of the contexts which may have led to the development of such collective assurance, namely the development of elite ties and relationships, before moving to more popular discourse. Again, this narrative has been informed by the retelling of the experience(s) of my respondents.

Elite relationships

> The United Kingdom has some important historical ties with Serbia [and Montenegro]. We were allies in the Second World War; indeed as the Foreign Minister pointed out to me, we were allies in the First World War as well. Our aim is that relations between the UK should resume their traditional closeness and that Serbia [and Montenegro] should become a leader in the region.

The quotation above is an extract dated 2005 (when Serbia and Montenegro were still unified) of the transcript of a press conference between then British Foreign Secretary, Jack Straw and Svetozar Marović, then president of Serbia and Montenegro (cited in Hodge, 2006: 197). This quotation, with its emphasis on Britain's history of friendship with Serbia, neatly summarises much of the positioning of parliamentary debate on the region. Over the last twenty years, concerning areas of the former Yugoslavia, debates have mainly focussed on: the conflicts in Croatia, Bosnia and Kosovo and in particular the role of British military personnel and international intervention; the arrangements made within a British context to provide for refugees fleeing those conflicts; post-conflict reconstruction; NATO intervention and the ongoing commitment of the British military as part of an international 'peace-keeping' force; refugee return and the issue of property restitution across the region; and moves towards EU membership. Observers have highlighted how debate at the parliamentary level around Britain's (non-)intervention in the conflict in Bosnia and later NATO involvement against Serbia in response to the conflict in Kosovo has been coloured by different theses (Simms, 2001; Hodge, 1999 and 2006). The 'ancient hatred' hypothesis (Dragović-Soso, 2008) emphasises historic ethnic antagonisms across the region which then translates into an insistence on moral equivalency, apportioning 'blame' to all sides equally. References to such moral equivalency dominate those parliamentary voices that repeatedly refer to the war in Bosnia as a 'civil war' and view the conflict as almost an inevitable part in a series of bloody episodes in the Balkans. Other common themes running through parliamentary debates around the conflicts are emphases on the apparent strength and stamina of the Serbs, thereby suggesting a military superiority which, by implication, should not be meddled with; the historic friendship between Britain and Serbia; the damaging effect attributed to 'early recognition' – particularly by Germany – of the independent state of Croatia; and dark hints at Russian-Serbian Orthodox kinship and at possible wider implications should Russia choose to assert a 'pan-Slavic' stance of comradeship with its Orthodox cousin:

> I hope that the Government have fully considered the effect of the bombing of Serbia on the struggling and fragile pro-Western and democratic forces in Russia. Pan-Slavism is a movement with deep historic roots and it is hard to envisage NATO's attacks on Serbia doing anything other than playing into the hands of the revanchist and former communist forces in Russia. The possible effect on Russia of the bombing in Serbia is more than worrying. We may yet be living dangerously all over again.[9]
>
> I want to sound a few warnings [...] I do not claim to be an expert on the Balkans but I am married to an Orthodox half-Russian. Over the years, I have come to know a little about the Russian soul and psyche. I give a solemn warning to the House and to the Government; they must not underestimate the intense pan-Orthodox and pan-Slavic feeling in Russia today. We may find it hard to understand, whether we are religious or not, but the Orthodox faith is a national faith of the Slavic people. We may be unleashing a tiger that we cannot control.[10]

Complementary to such references to the 'soul and psyche' of certain groups are unsubtle allusions to entire ethnic groups which, at times, make for uncomfortable reading:

> What Croat did to Slovene, what Slovene did to Serb, what they did to children was not new. It will not stop. It will not be stopped by the

words of western politicians. It is something that is built into the South Slav mentality.[11]

The war cannot be won; peace cannot be imposed; it is a tragedy deeply embedded in history that should be allowed to continue to unfold. Of course the Serbs have committed and are committing appalling atrocities, but so have the Muslims and the Croats in the past. Indeed, it is well known that the Croats and the Muslims, organised by the Nazis, murdered 500,000 Serbs during the war years and Serbian memories of that are still vivid. In fact, the west has consistently underestimated the Serbs. They are one of the fiercest, bravest and most patriotic race, and always have been.[12]

The Albanians are highly intelligent people. I speak of the whole Albanian race. They have a high level of literacy and a thirst for information.[13]

There is a wealth of evidence of Serbian toughness going back centuries.[14]

[The people of the Balkans] have the only European memory of a non-European imperial domination in a thousand years. We should have understood that a history like that produces not only cruelty and unreason but also a stubborn and reckless courage which is unknown to the rest of us.[15]

Highlighting the times of day devoted within British parliament to debates on the unfolding conflicts in Croatia and then Bosnia – usually in the small hours of the morning – as indicative of the priority given to debating the developing crisis by the British government, Simms (2001) observes how a relatively small number of Serb lobbyists shaped and influenced proceedings. This is a view reinforced by Hodge (1999 and 2006) in her analysis of the lobbying activities of British Serbs. MP Robert Wareing, who was suspended in July 1997 for failing to declare payments made to his consultancy firm by a Serbian company,[16] is particularly well-known for his pro-Serb position, particularly during the Kosovo conflict. Referring to the taking of UN peacekeepers as hostages during the Bosnian war as 'very bad public relations on the part of the Bosnian Serbs',[17] Wareing even six years after the European Community had formally recognised Bosnia-Herzegovina as an independent state, stated:

Bosnia is not a state or a nation, but a geographical expression. It consists of Serbs, Croats and Muslims, and even the Muslims are Serbs, because they are descendants of Serbs who converted to Islam during Turkish times. There is no such thing as a Bosnian. One can adopt that fiction if one likes but it is not likely to last.[18]

Robert Wareing along with MPs, Alice Mahon,[19] who acted as a defence witness in the trial of Milošević, John Randall and Daniel Kawczynski are vociferous in their assertions that the Serbs have been unfairly blamed for all atrocities committed in the Balkans and that acts of violence committed by non-Serbs have not been subject to the same judicial processes.

> Early in the break-up of Yugoslavia, the United States and some other NATO countries chose to demonise one of the groups of people who were suffering [...] There was no moral indignation in the media or from western leaders when 300,000 Serbs in Krajina were driven out of their homes by Croatian forces [...] During the past four years nothing has changed. The Serbs as a people are demonised in an increasingly racist way.[20]
>
> I take it from what the Minister is saying that he will be following up Interpol reports into organised crime and the links with terrorism in Kosovo. They contain a great deal of information. We should be seen as being even handed if we want people to go to The Hague.[21]

There are also however the voices of those who criticise the constant references to the conflict in Bosnia as a 'civil war' and the implications of moral equivalency in the atrocities committed:

> We must also be clear that this is not a civil war. We should not use the term 'warring factions'. How can a legitimate, internationally recognised Government be a warring faction? It beggars belief that anyone can attempt to classify in the same category Haris Silajdžić, the Prime Minister of Bosnia, and Karadžić, the leader of a minority of Serbs.[22]
>
> I am very glad that the Leader of the Opposition recognised in his speech the true origins of the horrific saga whose consequences we are discussing today. As he said, those consequences lie in the vicious dream of greater Serbian expansionism, and with those Serbians and their sinister intellectual backers who dreamed up the idea of smashing the Yugoslav federation and of turning areas of relative peace – contrary to the general mythology, families and villages had lived in peace for many years – into areas of ethnic hatred and horror. We should never forget that when we try to analyse how on earth we should wind down the spiral of hatred which has been escalated so viciously by those who are bent on ethnic cleansing.[23]

An Early Day Motion (EDM)[24] tabled by MP John Austin-Walker in January 1993 also expressed concern at the 'frequent references to the conflict in Bosnia as a civil war when it clearly results from external aggression from Serbia and Croatia against an internationally recognised sovereign state'.[25]

As part of informing the parliamentary debate around events in the region in the 1990s, researchers within the House of Commons library produced a number of reports.[26] Such reports tend to focus on a discussion of the military involvement of Britain and other NATO members and diplomatic negotiations and provide scant coverage of conditions on the ground for civilians. The April 1994 report for example (Watson *et al.*, 1994: 30) was written two years into the siege of Sarajevo, 20 months after journalists broke the news of the existence of concentrations camps in Bosnia (Vulliamy, 2012) and at a time when very few parts of Bosnia were untouched by atrocities committed against civilians. House of Commons researchers in that report discussing the question of British military intervention included the infamous quote from British General Michael Rose: 'the international community is not going to go to war over one broken down tank which is on its way out anyway'. The lack of attention paid in this series of House of Commons reports, written to inform MPs, towards the realities on the ground in Bosnia is particularly stark in the October 1995 (Watson, 1995) research paper which, in its three-page analysis of the fall of Srebrenica and Žepa, makes no reference to any atrocities committed nor to any massacre of civilians. In digesting those three pages, an uninformed reader would be forgiven for thinking that the shelling of Srebrenica had been purely a military exercise.

Elites and lobbying

The history of migration of Serbs to Britain and levels of collective assurance may go part way to explaining the greater sophistication in political activity carried out by British Serbs. I have already referred to Hodge's work (1999 and 2006) which analyses in painstaking detail the lobbying of British Serbs in the 1990s. In terms of direct, overt lobbying of members of British parliament, the clearest examples I found were on the part of Serb community organisations.

The Committee for Peace in the Balkans is an organisation chaired by Labour MP, Alice Mahon, referred to above. The name of such an organisation would imply activity relating to the conflicts in the region in the 1990s. And yet its main goal appears to be have been based around campaigning against NATO involvement in Serbia and the opposition of sanctions.[27] A prolific primary sponsor of parliamentary EDMs, it is interesting to explore the parliamentary activity of Alice Mahon related to the Balkans. It is possible to view all EDMs by primary sponsor dating as far back as the

1989–1990 parliamentary session. During the sessions from 1991 to 1996, Alice Mahon was primary sponsor for a total of 74 EDMs, not one of which related to the conflicts in the region. Of Alice Mahon's four EDMs as primary sponsor in the later 1998–1999 parliamentary session, three related to expressing concern at the effects of the bombing of Yugoslavia. The fourth related to the treatment of Serbs in Croatia and goes into surprising levels of localised detail into the 'blatant discrimination shown against Serbs [...] by the Government of Croatia [...] which is deliberately creating an ethnically pure state in Croatia'.[28]

The historic friendship between Britain and Serbia may have suffered a set-back in the (eventual) intervention on the part of NATO in the conflict in Bosnia and later in Kosovo but Serb community groups in Britain have been making concerted efforts to re-establish those ties. The following quotation is from a contribution to a parliamentary debate on Kosovo in 2008 by MP John Randall, described by representatives from Serb community groups as a 'friend of Serbia':

> I say to the British Government that we have a duty always to keep aware of what is going on in that part of the world [...] We have to give assurances to those people, regardless of who they are, that we in Britain – a traditional good friend of Serbia – and in the EU will look after them, and ensure, almost to repeat the words of the dictator Milošević, that we say, 'You shall not be beaten again'[29]

The symbolism of those words by Milošević, paraphrased by Randall, is well known as the moment in April 1987 when Milošević addressed a crowd of Serbs in Kosovo who felt threatened by their Kosovar neighbours, stating: 'No-one shall dare to beat you again'. Those words, interpreted as a virtual call to arms, have been described by Allan Little as 'the single sentence [which] changes his [Milošević's] life' and it is striking to hear them being voiced by a British parliamentarian twenty years later in encouraging intervention on the part of Serbs in Kosovo.[30]

The level of political sophistication demonstrated by Serb community groups was not observed in other diasporic groups. Any political work carried out by community organisations of other ethnicities appears almost naïve by comparison, which is not lost on the individual migrants:

> Yeah, we haven't been very good at all that PR type-stuff. The Serbs have had the Tories in their pockets for years. It's changing now, I think Arminka Helić has made a difference to some of the political dynamic.[31]

> (Bosnian interviewee, male, age 39)

There is evidence of more extensive and organised community-level political activity on the part of Serb groups in Britain than has been suggested for example by Pryke (2003) which, when set in historical context, could be considered to be both facilitated by and as a means of reinforcing Britain's historic friendship with Serbia.

Royal relationships

Relations between Britain and Serbia at the level of the state have also been reinforced via historical ties between the British monarchy and Serbian royal family in ways which have not been afforded in other states of the former Yugoslavia (being monarch-less). There are close ties going back decades between the Karađorđevićes, the Serbian royal family, and the British monarchy. In 1941, following the signing of the Tripartite Pact by Prince Paul of Yugoslavia, a coup d'état was instigated, exiling Paul Karađorđević to house arrest by the British in Kenya. Peter Karađorđević, who had until then been deemed too young to take the regency, at 17 was declared 'of age'. However, following Germany's full-scale invasion of Yugoslavia in April, 1941, Peter II and his government, who had already been educated in England, went into exile in the US, Cairo and London. Peter II completed his education at the University of Cambridge and went on to serve in the Royal Air Force. Deposed by the Yugoslav Communists following the end of the war, Peter had a son, Crown Prince Alexander II, at the time of writing the would-be monarch of Serbia, born in suite 212 of Claridge's hotel, London, declared as Yugoslav territory for the day by Churchill. Alexander and his family's Yugoslav citizenship was revoked by the Yugoslav authorities and their property was confiscated. The family's first visit to Yugoslavia following the period in exile was in 1991 and they were conferred citizenship by the Federal Republic of Yugoslavia in 2001. Crown Prince Alexander II was baptised at Westminster Abbey by the Serbian Patriarch with godparents King George VI and his daughter, Princess Elizabeth, the future Queen Elizabeth II in attendance. Like his father, he was educated at schools in England and served for a time in the British military. By marrying a catholic, Crown Prince Alexander II lost his place in line of succession to the British throne, which he held by virtue of being a descendent of Queen Victoria. His three sons remain in line of succession to the British throne, were all educated at schools in London and Canterbury and two went on to study at universities in London. Crown Prince Alexander II's paternal grandmother, Queen Maria of Yugoslavia remained in England after the family's exile. She died in London in 1961 and was buried in the royal burial ground adjoining Windsor Castle. In 2013 the Karađorđević family repatriated the remains of several of their family members to be buried in a state funeral at the Serbian Royal Mausoleum at Oplenac. In west London one month before the state funeral in Serbia, a memorial service was held for Queen Maria at the Serbian Orthodox Church St Sava. Speaking in English, Crown Prince Alexander expressed his gratitude to the country that had granted his family exile and was emotional when speaking of his grandmother's return 'home'.

Another royal link between the House of Karađorđević and the House of Windsor existed in the relationship between Princess Margaret, sister of Queen Elizabeth II and Prince Nikola of Yugoslavia, the son of the Regent Prince Paul, Peter II's uncle whose rule had been overthrown by the Allies when Peter II was declared of age. Prince Nikola was educated at the University of Oxford and there was speculation in the press at the time about a possible marriage between the Prince and Princess Margaret. He was killed in a car accident in Buckinghamshire in 1954. Nikola's older brother Prince Alexander is one of the four founding members of the Serbian Unity Congress (SUC), a Serbian diaspora international organisation with its main office based in Washington and national chapters in several other countries. The SUC have submitted evidence to the British parliament on a number of occasions.

The previous section has summarised some of the discourse around the former Yugoslavia from the perspective of state-level institutions. I now move on to considering how the region is represented through more popular discourse, namely the media, another facet of the general context of reception into which migrants arrive, are making their daily lives and can contribute to the formation of relationships both with the country of origin and country of settlement.

Media discourses

Serbs in Britain have been described by one journalist as a 'community besieged by journalism's snipers'.[32] Serbophile MP Robert Wareing, some of whose contributions to parliamentary debates on the former Yugoslavia have been discussed earlier in this chapter, was unequivocal on his perceptions of the portrayal of Serbs in the media:

> We have to look into the reality of the situation, and into the minds of many Serbs, [who] in Serbia have been living for the past eight years against a background of anti-Serb hysteria in the media, particularly in the west – in America, Britain and Germany – which has demonised every Serb, saying that all of them were evil. The hysteria has been deeply regrettable.[33]

Interviews with respondents showed perceptions of bias, irrespective of country of origin or ethnicity, illustrated by the following quotes:

Interview with Bosnian migrant:

The Serbs had virtual control of their media representation in the UK. It was clear to me who was in control and let's just say, it wasn't Bosnians.

Interview with Croatian community representative:

The propaganda here during the war from the Serb side was heavily biased. We would ring each other up and say, have you heard the news, they're saying Dubrovnik was being shelled not by Serbs but by Croats, etc.

Interview with Serbian migrant:

I was very upset, the media in the West [during the conflict in Bosnia] was very Bosniak, it was so transparent.

These three quotes highlight how diasporas representing different ethnicities have perceived levels of bias in the reporting of the conflicts in Bosnia and Croatia in Britain. Articulations of unjust victimhood and narratives of unfairly attributed blame run through any references to the media in participant responses. In some senses, the *actual* reporting in the media is probably of less importance for any analysis of transnationalism than the *perception* of that reporting. Transnational characters and affiliations are affected by the ongoing discourse of the context of reception and it is how that discourse is perceived by migrants that is key to creating that dynamic, rather than an external review of such discourse. We can see how that dynamic is shaped through the narratives of Serb respondents who perceive that they are victimised by 'Western' media versus non-Serb narratives who feel as though the media is unfairly pro-Serb. What is actually reported tends to get lost amongst the rhetoric.

It is possible to see from the kinds of publications[34] that feature articles including keywords of Slovenia and Croatia how their accession to the European Union and discussion of economic conditions have dominated media discourses around these two particular former Yugoslav states. Slovenia and Croatia's accession to the EU has led to their inclusion in comparative studies of living conditions across EU member states, which in turn has led to the development of a theme of focus within British media that I refer to as the 'even Slovenia' theme. Generally covered outside of the broadsheets, these reports of Europe-wide studies show various aspects of life in Britain as being 'behind' Croatia and Slovenia in a number of EU standard of life assessment reports.[35]

The ways in which the British print media have reported the conflicts in the region have been the subject of study by Kent (2006); Simms (2001) and Robison (2004). The tabloids, and to a lesser extent the broadsheets, focussed on Britain's role as the provider of sanctuary for those fleeing the conflict seeking refuge. Using terms such as 'mercy missions', details were provided of medical evacuations of the wounded, often children, from the conflict areas, to hospitals in Britain. Steve Crawshaw in the *Independent* condemns Britain for congratulating herself on its magnanimous and generous support of refugees from the region when in real terms and proportionate to the population Britain was 'almost at the bottom of the list' in the numbers of those given sanctuary. Crawshaw describes the high-profile evacuation operations of limited numbers of telegenic children as a 'cynical operation […] a high-profile side-show – distracting the audience with a set of bleeding-heart fireworks, while the sordid business is wrapped up elsewhere'.[36]

The *Independent* also commented on Bosnian refugees' experiences of the British asylum system, which highlighted the different conditions placed upon those who travelled to the UK as part of coordinated refugee agency projects and those who made the journey independently.[37] The *Guardian* draws attention to the case of one Bosnian man who had been granted asylum in the UK, after being interned in three different Serb concentration camps, but whose family members were raped and killed in Bosnia following Home Office rules on the definition of 'dependents' under family reunification policies.[38] The same year, the *Guardian* reports on 'buck-passing' between the UNHCR and the Home Office resulting in significant delays in reuniting those men who had been detained in camps in Bosnia, granted asylum in the UK with their wives and children still in Bosnia.[39] And earlier in the conflict Britain had been accused of failing to accept their 'share' of refugees.[40] The distressing experience of refugees, from Bosnia and Kosovo, when they arrived in Britain was also covered in some detail.[41]

I have discussed earlier in this chapter how interpretations of the conflicts in the Balkans – particularly in Bosnia – were dominated in British parliamentary discourse by the 'ancient hatred' hypothesis (ÓTuatheil, 2002). Similar discourses were presented via some of the British print media. The following article is one example of the kind of narrative in the popular press that would have welcomed arrivals from Bosnia in 1995. The *Daily Mail* cited former Minister for Defence Alan Clark in May 1995 who lingered on the events of WWI and WWII in making the case for non-intervention in the Bosnian conflict. At the time Clark was writing, the situation in 'safe haven'

Srebrenica in eastern Bosnia had deteriorated to the point of catastrophe for its Bosnian Muslim residents. By June 1995 Bosnian civilians in the enclave were dying of starvation. In July 1995 more than 8,000 Muslim men and boys of Srebrenica were separated from the women and girls of the area and killed by units of the VRS under the command of Ratko Mladić. Muslim women and girls who had taken refuge in the nearby village of Potočari were raped by Bosnian Serb forces with witnesses reporting testimony of babies and young children being forcibly removed from their mothers and killed and male and female civilians being subject to acts of torture and rape within sight of Dutch UN forces. The ICTY has found the massacre at Srebrenica to be an act of genocide. Yet, Alan Clark, in the words of the *Daily Mail* a 'leading military historian', describes Bosnian Serb para-militaries as a 'pretty rough lot', concluding his argument: 'it must be said that what is happening in the former Yugoslavia offers a classic example of the fate that awaits all federations imposed on proud and independent races who speak with differing tongues and have long histories of mutual antagonism'.[42]

The position of the broadsheets to the continuing conflicts was more ambiguous. Brendan Simms (2001) in his account of the British establishment's policy failures over the Bosnian conflict lays out what he sees as the editorial line of each of the broadsheets with the *Sunday Telegraph*, the *Sunday Times* and *The Times* being credited as supporters of 'limited military intervention' to contain the Bosnian Serbs, the *Independent* and the *Daily Telegraph* charged with 'meandering back and forth' and the *Guardian* and the *Observer* being 'consistent in their refusal to countenance military action on Bosnia's behalf or even to allow the Bosnians to defend themselves' (Simms, 2001: 303). Simms does distinguish however between the editorial line of the broadsheets and the positions of individual journalists who 'told the story more or less as it was [who] uncovered the camps, they highlighted the misery of Mostar and Sarajevo and they helped to expose the truths on which western policies were founded' (ibid.: 301).

I have presented the above examples as some of the discourse and context of reception into which migrants from the former Yugoslavia would have been arriving. Such discourse can affect the levels of security and 'sense of self' that a migrant develops within the host state, impacting upon the transnational dynamic. Before moving to the following section on transnational triggers, I offer some reflections on the ways in which perceptions of media portrayals of Serbdom may have influenced the development of the transnational space of both Serbs and non-Serbs alike.

The question of whether public discourse around the former Yugoslavia in Britain is dominated by Serbophilia or Serbophobia has been the subject of debate, with the voices of those claiming Serbophilia (Simms, 2001;

Hodge, 1999, 2006) pitted against the voices of Serb communities in Britain who point to examples where they believe they have been victimised, particularly in the press.[43] The positive discourse around British–Serbian relations prior to the conflicts in the 1990s is a constant reminder to British Serbs of a relationship which many see as spoiled especially since Britain's role in NATO's intervention. As early as 1993 the *Guardian* was reporting on Britain's policy in the conflict in Bosnia and its position on sanctions against Croatia as being 'pro-Serb' and particularly damaging for the situation of Bosnian Muslims. One *Guardian* article quotes Austria's *Die Presse* which argues that 'The motive of Her Majesty's government is quite transparent. London is charged with co-responsibility for the Balkan catastrophe because it has long held a protective hand over the Serbs'.[44] This evocation of a protective hand contrasts sharply with the perception of some Serbs later in the conflict following air-strikes against Serb-held positions in Bosnia and certainly later that decade with the British government advocating for NATO intervention in Serbia in response to Serbia's actions against the Albanians of Kosovo.

It is also interesting to consider this notion of a 'protective hand' within a historical context. As WWI unfolded, the British press were quoted as proclaiming:

> There has been nothing more thrilling or heroic in the whole war than the extraordinary victory won by the Serbs when they seemed spent to prostration. [...] We were in less sympathy with them [the Serbs] as with others after the Second Balkan War but we take back our doubts. They have proved themselves worthy of a greater future and they are sure in the end to achieve it.[45]

And the Serbian Prime Minister in 1916 was quoted as saying:

> Never can we be thankful enough to England and our other Allies for the generous and unlimited help they have given us. Never can we forget that during the period when our press was silenced, owing to the enemy invasion, it was your press which defended us so nobly, and placed our position before the eyes of the world.[46]

This mutually supportive relationship was seen to crumble only towards the end of the Bosnian conflict and then took a more substantial body blow in Britain's intervention in Kosovo. One Serb refugee from Krajina is quoted in August 1995 as saying 'You British were our allies in every war before. But you've betrayed us in this one.'[47] One way of interpreting the political lobbying on the part of Serbs in Britain would be an attempt to re-establish old British-Serbian friendships. In this sense the transnational dynamic is a purposeful, active space characterised by a clear goal and commitment, aims which can sometimes be crystallised by political 'triggers'.

Transnational 'triggers'

Throughout the course of the research it became evident that there were certain events which acted as identifiable catalysts to alter the trajectory of a migrant's expressed transnational being and belonging. Several interviewees articulated their emotional responses around the high-profile arrest and extradition of those indicted for war crimes, with some describing how media coverage of such events generated intense memories and even flashbacks.

In response to action on the part of Serbian forces in Kosovo in 1999, NATO carried out Operation Allied Force, a 78-day campaign of air bombardment, ostensibly against military targets within the territory of the Federal Republic of Yugoslavia. The operation was particularly controversial as this was the first time that NATO had employed military action without the endorsement of the UN Security Council. The air campaign attracted much criticism particularly for its use of cluster bombs and its strikes against civilian targets, including hospitals and schools. The bombardment was identified by a number of Serb respondents as a key pivot around which their strength of ethnic identity intensified. Anger around the position that many felt they had been placed in as British nationals and voters (some carrying dual citizenship) grappling with visions via the media of their homeland being bombarded by NATO, coupled with worry about family members and friends still resident in Serbia, formed a distinct thread through the narratives of almost all Serb interviewees and survey respondents. One writer on the British-Serb community website, *Britić,* described how the NATO bombardment had 'changed his entire life' in terms of his identity formation.[48] Ivana talked about how she had rejected her Serb ethnicity during her first few years in Britain, hiding her identity for fear of being branded a nationalist or a war criminal until NATO action propelled her to embrace her ethnic background, thus becoming significantly more defiant in expressing her ethnic identity to others:

> To be honest that singular event changed everything for me. Before that I was so ashamed, if anyone asked me where I was from I always said Yugoslavia. Meeting new people was always a bit awkward, especially during the 1990s when we were total pariahs. Then when that happened, I couldn't believe it. How could that happen to a European

state? From that point on, I was not going to hide away anymore. I am Serbian and everyone is going to know it. And if they have a problem with that then they can go and walk off a bridge. At least you have bridges. Ours are all bombed.

(Serbian interviewee, female, age 36)

It is also interesting to compare the experiences of those migrants who came at different points in time. The Slovenian and Kosovar embassies for example, hold parties and events aimed at their diasporas and appear to welcome with open arms those registered. Compare this with the experience of those who arrived in Britain following WWII and were unable, or unwilling to return for fear of being arrested as 'dissidents'. Serb respondent, Branko, who came with his parents as a child and whose father was one of the founding members of the Serbian community club in London remembers being physically ejected from the Yugoslav Embassy as a child with his mother. And yet, his mother, who only returned to Serbia more than 30 years after she had first left and her daughter, who was born in England, cherish their Serb identities, regardless, or perhaps in part because, of how they were treated at the hands of Serbian authorities.

This chapter has sought to contextualise some of the different discourses into which migrants from different parts of the former Yugoslavia would have been arriving and making new lives. The history of migration by Serbs to Britain and the development of elite relationships has been highlighted as a means of demonstrating how such contexts can influence the transnational dynamic. The heterogeneity of the experience of migration from the region is clear through the articulations of my respondents. The challenges of negotiating the British immigration system – regardless of country of origin or 'type of migrant' – were also reflected across the body of narratives as were the consequences of living 'in limbo'.

Notes

1   The term used by the writers of the report.
2   One of the Serb community representatives indicated that they estimated there to be up to 100,000 Serbs living in Britain.
3   Seton-Watson archive, SSEES, box SEW/7/3.
4   In response to the labour shortage in the post-WWII era, the British government recruited workers from European countries into (usually) industrial posts throughout Britain.
5   Interview with Serbian community representative, April 2008.
6   Such 'look and see' programmes eventually developed into the Assisted Voluntary Return Programmes similar to those operated by the International Organisation for Migration (IOM) and Refugee Action which offered financial and other incentives to returning refugees.

7 In 2003 it was estimated that of the 4,346 Kosovar evacuees to the UK under the Humanitarian Protection Programme 500 remained in the UK (Smart, 2004: 26).

8 Interview recorded for the Evelyn Oldfield Unit Refugee Communities History Project and archived at the Museum of London (reference 2005.160).

9 Lords Hansard transcript for 6 May 1999.

10 House of Commons Hansard Debates for 25 March 1999.

11 Lords Hansard transcript for 6 May 1999.

12 House of Commons Hansard Debates for 31 May 1995.

13 Lords Hansard transcript for 6 May 1999.

14 Ibid.

15 Ibid.

16 'Labour suspends MP in Lobbying Row, *Independent*, 19 June 1997.

17 House of Commons Hansard Debates for 31 May 1995.

18 House of Commons Hansard Debates for 6 December 2001.

19 MP for Halifax, home to a significant Serb population and nicknamed by one journalist as a 'Serbian enclave in Yorkshire' ('A Serbian enclave in Yorkshire', *Independent*, 26 July 1995).

20 House of Commons Hansard Debates for 17 June 1999.

21 House of Commons Hansard Debates for 6 December 2001.

22 House of Commons Hansard Debates for 31 May 1995.

23 House of Commons Hansard Debates for 31 May 1995.

24 EDMS are formal motions submitted by MPs for debate in Parliament, usually used to highlight a particular issue of concern to the MP or his/her constituents. EDMs are introduced by a 'primary sponsor' with MPs adding their signatures to those EDMs they wish to support.

25 Early Day Motion 1,203 – Aggression against Bosnia-Herzegovina.

26 Watson *et al.* (1994); Watson and Ware (1994); Watson (1995); Watson and Dodd (1995); Ware *et al.* (1995).

27 'Anglo-Yugoslav Medical Aid Newsletter, June 2000', Imperial War Museum archive K 00 / 1735; NATO protest posters, Imperial War Museum archive, PB 71/7 (497–1).

28 Early Day Motion 1,000, 19 November 1999.

29 House of Commons Hansard Debates for 28 February 2008.

30 A. Little, 'Slobodan Milosevic's road to ruin', BBC News. http://news.bbc.co.uk/1/hi/world/europe/4819388.stm (accessed 22 April 2016).

31 Arminka Helić, special advisor and Chief of Staff to Foreign Minister and Conservative politician William Hague, described by the *Daily Mail* as 'Hague's blue-eyed Bosnian Muslim émigré'.

32 'A Serbian enclave in Yorkshire', *Independent*, 26 July 1995.

33 House of Commons Hansard Debates for 17 June 1999.

34 For example, *Euroweek*.

35 'Even Croatia beats Britain on trains', *The Evening Standard*, 10 January 2002; 'Britain's pension shame: UK state pensions are the LOWEST in Europe – and even Slovenia and Slovakia give their old people more', *MailOnline*, 27 November 2013; 'Britain's cancer shame: UK has lower survival rate for cervical cases than Slovenia and Czech Republic', *MailOnline*, 21 November 2013; 'Why Britain's no place to be a mum', 7 May 2013; 'Slovenia beats Britain in safe births league', *The Times*, 18 October 2007; 'Why it's better to grow up in Slovenia than Britain', *Daily Mail*, 10 April 2013; 'Slovenia is now a better place to grow up than Britain', *The Telegraph*, 10 April 2013; 'Quit Britain,

you'll be better off in Slovenia', *Daily Star*, 31 October 2013; 'Britain trails even Slovenia in state school class sizes', *The Daily Telegraph*, 19 September 2007.

36   '"A delayed victory by Hitler": Britain's response to Bosnia has uneasy resonances for Germans', *Independent*, 18 August 1993.

37   'Bosnia victims cannot stay in Britain: War refugees are being kept in limbo over their future', *Independent*, 25 August 1993.

38   'Refugees "betrayed" by inflexible rules', *Guardian*, 31 May 1993.

39   'Forgotten; freed from Serbian camps and settled all across Europe, ex-detainees from Bosnia find their troubles are not over', *Guardian*, 7 May 1993.

40   'Britain attacked for ignoring Bosnia refugees', *Independent*, 27 July 1992.

41   '"I believed in justice": Refugee week', *Independent Extra*, 21 June 2007.

42   'Britain must pull out of Bosnia now', *Daily Mail*, 30 May 1995.

43   'One in five British Serbs "victimised" or "threatened"': www.ebritic.com/? p=199216 (accessed 22 April 2016).

44   'Villains of the peace in the Balkans', *Guardian*, 2 September 1993.

45   'Heroic Serbia', *The Observer,* 13 December 1914.

46   'The Voice of Serbia: An interview with M. Pasitch', *The Manchester Guardian*, 3 April 1916.

47   'When Britons rallied to "heroic Serbia": Serbs regard Britain as a historic ally and cannot understand why they are now treated as pariahs', *Independent*, 13 August 1995.

48   www.ebritic.com/?p=206581 (accessed 22 April 2016).

References

Bajić-Hajduković, I. (2008) 'Belgrade parents and their migrant children', PhD thesis, University College London.

Bloemraad, I. (2004) 'Who claims dual citizenship? The limits of postnationalism, the possibilities of transnationalism and the persistence of traditional citizenship', *International Migration Review* 38(2): 389–426.

Collyer, M., Binaisa, N., Qureshi, K., McLean Hilker, L., Oeppen, C., Vullnetari, J. and Zeitlyn, B. (2011) 'The impact of overseas conflict on UK communities'. York: Joseph Rowntree Foundation.

Dragović-Soso, J. (2008) 'Why did Yugoslavia disintegrate? An overview of contending explanations', in L.J. Cohen and J. Dragović-Soso (eds) *State Collapse in South-Eastern Europe: New Perspectives on Yugoslavia's Disintegration*. West Lafayette: Purdue University Press.

Hodge, C. (1999) *The Serb Lobby in the United Kingdom*. Donald Treadgold Series, No. 22. University of Washington: Seattle.

Hodge, C. (2006) *Britain and The Balkans: 1991 Until Present*. London: Routledge.

Jančić, M. (1996) *The Flying Bosnian: Poems from Limbo*. London: Hearing Eye.

Jarowsky, B., Levitt, P., Cadge, W., Hejtmanek, J. and Curran, S.R. (2012) 'New perspectives on immigrant contexts of reception: The cultural armature of cities', *Nordic Journal of Migration Research* 2(1): 78–88.

Kelly, L. (2003) 'Bosnian Refugees in Britain: Questioning Community', *Sociology* 37(1): 35–49.

Kent, G. (2006) *Framing War and Genocide*. Cresskill, NJ: Hampton Press.

Kershen, A. (ed.) (1997) London, the promised land? *The Migrant Experience in a Capital City*. Aldershot: Ashgate.

King, J. (2001) 'Factors affecting Australia's refugee policy: The case of the Kosovars', *International Migration* 39(2): 73–92.

Kjaerum, M. (1994) 'Temporary protection in Europe in the 1990s', *International Journal of Refugee Law* 6(3): 444–456.

Koser, K. and Black, R. (1999) 'Limits to harmonization: The "Temporary Protection" of refugees in the European Union', *International Migration* 37(3): 521–543.

Kostovicova, D. and Prestreshi, A. (2003) 'Education, gender and religion: Identity transformations among Kosovo Albanians in London', *Journal of Ethnic and Migration Studies* 29(6): 1079–1096.

Mares, P. (2001) *Borderline: Australia's Treatment of Refugees and Asylum Seekers*. Sydney: UNSW Press.

Mavra, L. (2010) 'Contesting migrant identifications: Community, friendship and ethnicity among London's Serbs', PhD thesis, Kings College London.

Mazzucato, V. (2007) 'The role of transnational networks and legal status in securing a living: Ghanaian migrants in The Netherlands', COMPAS working paper no. 43, University of Oxford.

ÓTuatheil, G. (2002) 'Theorizing practical geopolitical reasoning: The case of the United States' response to the war in Bosnia', *Political Geography* 21(5): 601–628.

Padin, José Antonio (2005) 'The normative mulattoes: The press, Latinos, and the racial climate on the moving immigration frontier', *Sociological Perspectives* 48: 49–75.

Portes, A. and Rumbaut, R. (2006) *Immigrant America: A Portrait*. Berkeley: University of California Press.

Pryke, S. (2003) 'British Serbs and long distance nationalism', *Ethnic and Racial Studies* 26(1): 152–172.

Robison, B. (2004) 'Putting Bosnia in its place: Critical geopolitics and the representation of Bosnia in the British print media', *Geopolitics* 9(2): 378–401.

Simms, B. (2001) *Unfinest Hour: Britain and the Destruction of Bosnia*. London: Penguin.

Smart, K. (2004) 'Refugee populations in the UK: Kosovars'. ICAR navigation guide.

Stepick, A. and Stepick, C.D. (2009) 'Diverse contexts of reception and feelings of belonging', *Forum: Qualitative Social Research* 10(3).

Vathi, Z. (2010) 'New Brits? Migration and settlement of Albanian-origin immigrants in London', working paper no. 57, University of Sussex: Sussex Centre for Migration Research.

Vulliamy, E. (2012) *The War is Dead, Long Live the War: Bosnia: The Reckoning*. London: Vintage.

Walsh, M., Black, R. and Koser, K. (1999) 'Repatriation from the European Union to Bosnia-Herzegovina: The role of information', in K. Koser and R. Black (eds) *The End of the Refugee Cycle? Refugee Repatriation and Reconstruction*. Oxford: Berghahn.

Ware, R., Watson, F. and Dodd, T. (1995) 'Bosnia: Update and supplementary paper'. London: House of Commons Library.

Watson, F. (1995) '"Not peace, but a big step forward": Bosnia in October 1995'. London: House of Commons Library.

Watson, F. and Ware, R. (1994) 'Bosnia, the UN and the NATO Ultimatum'. London: House of Commons Library.

Watson, F. and Dodd, T. (1995) 'Bosnia and Croatia: The conflict continues'. London: House of Commons Library.

Watson, F., Dodd, T. and Ware, R. (1994) 'Bosnia: The "Sarajevo Formula" extended'. London: House of Commons Library.

3   The lexicon of the migration experience

The ways in which the lexicon around migration is adapted and manipulated within the popular discourse and (anti-) immigration rhetoric has been and continues to be the subject of much contested debate at a time when migration crises are hitting the headlines. The desperate situations of those seeking refuge from violence, persecution and economic crisis by attempting to cross the increasingly fortress-like 'key' borders around and within Europe have, at the time of writing, reached zenith proportions of tolerability. The conditions under which those attempting to seek asylum in different European states are being forced to make their journeys have been described as like 'living in a horror movie'[1] and call into question the whole etymology of the word 'refuge'. Turton (2003: 2) argues that 'we need to become self-conscious' about the use of language associated with migration. Certainly, in the often highly charged and politicised narrative around (im)migration in the British context, the language ascribed to the migration process and those who migrate is constantly being renegotiated as images and symbols associated with such language are formed in the public imagination. The lexicon of migration also becomes increasingly contested as the struggle between those who wish to restrict the flows of people into the country and those who argue for the rights of human beings to migrate intensifies.[2]

Considering all the possible connotations associated with the various terms that can be used to describe the status of someone who moves their home from one place to another, I made the decision, throughout my research, to settle on what I thought of as the relatively value-free term (at least when compared with the other options) of 'migrant'.[3] I came to realise however that whilst the term of migrant, as I had conceptualised it, may be the most neutral of the different possibilities , thereby affording the most open and inclusive way of recruiting potential participants, a term that *I* might have considered to be 'value-free', it was certainly not interpreted as such by my respondents.

Levitt and Glick Schiller (2004) use the concepts of 'being and belonging' to articulate how memory and the imagination affect the transnational relationship. Whilst the impact of physical distance on emotions and the subsequent influence over family dynamics is one aspect of the debate around transnational emotions, the territorial dimension is part of a complex and multi-layered labyrinth of ties and motivations connecting those in the diaspora with those they have 'left behind'. The multitude of different articulations of emotions as expressed by migrants from the former Yugoslavia in Britain during the course of my research provided an insight into the intricacy of the different emotional connections, ties, motivations for engagement in the transnational space and triggers of those emotions. Amongst the 'bank' of emotions as expressed through the course of the transnational narratives of migrants arose articulations (including denials) of love, loss, anxiety, sadness, guilt, loyalty, bitterness, pride, jealousy, embarrassment, relief, nostalgia, frustration, happiness, disappointment, *domoljublje*. This last emotion, *domoljublje*, one possible translation of which is 'patriotism', can etymologically be translated as 'love of one's home'. It is interesting to unpick this term and concept further in relation to those articulated sentiments towards the home and homeland as expressed by participants in the project.

Different languages express 'love of one's homeland' in different ways and the English 'patriotism' misses certain dimensions of the intensity of feeling that one who has left his/her home can have with the homeland. Some languages have a word to express love of one's homeland which encompass a different breadth of concept than implied by 'patriotism', perhaps reflective of that particular country's history of outward migration or enforced denial of certain aspects of its culture. The Welsh word *hiraeth* for example has been defined as 'homesickness tinged with grief or sadness over the lost or departed. It is a mix of longing, yearning, nostalgia, wistfulness, and the earnest desire for the Wales of the past'[4]. The term *domoljube* in Bosnian/Croatian/Serbian could of course apply not just to the homeland as a nation-state but could equally apply to the more local *zavičaj* or also to the family unit (Halilovich, 2013). We can also see, however, expressions in the narratives of some migrants – particularly in those who did not 'choose' to leave – that would suggest a level of intensity of feeling toward the 'home' more reflective of the emotion expressed by the concept of *hiraeth*, as opposed to 'patriotism'. This intensity then adds another dimension to any transnational space especially when combined with other more ambiguous feelings related to experience(s) of trauma.

Whilst familial transnational emotions played an important role in the articulation of transnational ties of migrants, it would be simplistic to focus solely on 'the family' as the lens through which all transnational emotions can be viewed. The diaspora can play both an allegorical and literal familial function through the narratives of some migrants with individual members and the diaspora as a collective being welcomed, celebrated, avoided, relied upon and, in some cases, representing dispute, such as may be seen through the dynamics of many families.

This chapter therefore discusses some of the ways in which the trans-national and diasporic space is moulded and shaped by the lexicon of the migration itself and its associated emotional connotations, connotations which form relationships between the migrants and his/her identity with the homeland, the new home and with others in the diaspora. I turn first of all to a discussion of how identity as a 'migrant' was articulated by my respondents.

Articulations of migration, exile and 'refugeehood'

The (not a) 'migrant'

Borrowing on the semiotic system as developed by Barthes and Saussure, the sign of <chosen term to represent the person who migrates> will be decoded differently by the receiver, depending on a whole host of contex-tual factors, such as the receiver's age, gender, cultural background, life experience, proficiency in the language being used and so on. What I have determined to be 'value-free' may therefore not be interpreted as such by someone with a different set of cultural references from mine. It is possible, through the narratives of my respondents, to discern how the words with which 'migrants' choose to describe themselves can form a core element of the articulation of their transnational sense of being and belonging. Whether someone identifies as a 'refugee' for example or an 'exile' or a 'migrant' can speak into the relationship that that individual has with the country of origin, with the new home or with others within the diaspora. Any transna-tional space(s) within which the migrant may be acting or 'simply' being are framed by the migration lexicon itself and can therefore change as the vocabulary changes.

The term of 'migrant' then, whilst possibly a relatively neutral term in the field of migration studies is interpreted by some in the light of juxtaposi-tion of 'migrant' with prefixes such as 'economic' and 'labour' (Berg, 2011: 22–23). Some of my respondents carried a similar mental juxtaposition of the term 'migrant' with 'economic'. We were introduced in Chapter 1 to Serbian Ana who came to Britain as a young woman in the 1960s. Ana prefaced the beginning of her interview with the comment: 'I am not who you want to speak to. Am not even a migrant – mine is a love story – it was love that brought me here.' Economic factors do not, at least in Ana's sphere of understanding, play a part in the narrative of her migration trajectory. She therefore does not consider the term of migrant, or indeed any of the terms that I had spent some considerable time grappling with, to apply to her experience at all – hers is not a story of migration at all, as she says, hers is a love story. In Ana's articulated identity as 'not a migrant', she is expressing a difference, separating her reasons for making the journey from those of 'others' who she perceives as coming to Britain for reasons not comparable with hers.

Interestingly, we can see through Ana's narrative other articulations and extensions of this 'otherness' from her diasporic peers: Ana has at different points during her life in Britain affiliated with Serb-organised community groups but does not consider herself to be a beneficiary or a recipient of services; hers is always the benefactor role of a 'mentor–mentee' relationship where, as one of the earlier (non-)migrants to have moved from Serbia to Britain, she acts as counsel and advisor to those making the journey at different points after her. Throughout her story, we see articulations of her desire to 'give something back' to her compatriots, both those she has left behind and those who she sees as following in her footsteps. I return to this later in the chapter when discussing narratives of guilt but we can see how Ana's articulation of her identity as a non-migrant is a reflection of her independence from what she perceives to be migrants in need of assistance. In Ana's transnational world therefore she holds herself from her diasporic 'others' – dispensing knowledge as one with more well-developed experience of life in Britain.

The 'refugee'

The term 'refugee' carries with it complex and multi-layered connotations and values. My interpretations of the term are in light of understandings as articulated by my respondents. Such articulations may not always correspond with the definition outlined in the 1951 Geneva Convention and much less with the Home Office's determination of who is to be granted the elusive 'refugee status'. Turton (2003) in reviewing the matrices of migration as proposed by Richmond (1994) and Van Hear (1998) argues that even amongst a detailed and complex continuum of 'forced', 'voluntary', 'reactive', 'proactive' migrations and everything in between, what is lacking across all the debates is the agency of individual migrants. I would suggest however that a concern over individual agency at the beginning of the migration process or at the point of departure risks a neglect of what is awaiting the (whatever type of) migrant when it comes to engaging with the often protracted processes of the immigration system in the destination country.[5] Some narratives of migrants from the former Yugoslavia reflect the institutional and bureaucratic nightmare that can represent the British immigration system, a system which can have the effect of creating black holes in which anything resembling agency, choice or free will on the part of individual migrants can gradually disintegrate over years of limbo and frustration.

It is interesting to consider some of the potential connotations as associated with the term 'refugee' and particularly the impact that local context can have on different interpretations. By way of example I offer four different contexts related to the former Yugoslavia:

1. Refugee as status holder and source of benevolence (Bosnia)

In 2002 when working in Sarajevo, I was sitting in a café with a local (Bosnian) friend. It was late summer and a Saturday night and the centre of Sarajevo was packed with people out socialising with friends, in restaurants and walking up and down *Ferhadija*, the pedestrianised thoroughfare in the centre of town, lined with shops, bars and cafes. Near the café was a side road which, usually empty, was full of parked large, top-of-the-range Mercedes cars with foreign number plates. It was quite a striking sight and I remarked to my friend that I didn't think I had seen so many large, impressive looking cars all parked in one place before. My friend's reaction was to look disinterested and dismiss the scene with a wave of the hand and a one-word comment: 'refugees'. Sarajevo and other towns in Bosnia are full on weekends of those who had left either during or after the war to either neighbouring states or other European states such as Austria or Switzerland and who return on weekends and during the holiday season to visit friends and family. The Mercedes and other trappings of wealth are status symbols as proof that they have 'made it' in their new homelands. This was reflected in the narrative of some of my interviewees who expressed their wish to somehow 'prove themselves' when 'over there'. One Kosovar male interviewee joked about the pressure he felt to portray a certain level of financial standing when he returned to Priština on visits: 'Sometimes after I come back (to England) I live on baked beans for a week or two. Because all my money has been spent on taking back bottles of single malt. I have a lot of uncles.' In this example the refugee is seen as a wealthy émigré showing a degree of benevolence and a desire to prove status through the accumulation of material possessions.

2. Refugee as disloyal deserter (Kosovo)

In 2004 I was visiting friends in Priština and discussing with one of my friends his plans for the future regarding work and where he was planning on living. This was a Kosovar Albanian man in his early 30s with a young family who, at the time of the conflicts in Kosovo, would have been in his mid-to-late 20s. Many of his peers had left Kosovo at that time and had chosen not to return. He spoke with some passion about the feelings of contempt that he had for those refugees who, in his opinion, had 'abandoned' their country at a time when it needed them most. He and his wife both worked for large international NGOs and he described his frustration and feelings of helplessness at what he perceived to be the necessity to entrust governance of his country to international organisations because so many of the young, talented and ambitious Kosovars had 'chosen' to leave. This discussion took place in a café in Priština where the electricity kept cutting out intermittently and where the whole café carried on their evening (except for the card players) even as the café was being plunged into darkness every 20 minutes or so. Despite the obvious problems that Kosovo was experiencing at the time (the lack of electricity being one), my friend was determined to remain living in his village an hour's drive away from Priština with his family and in his view to play a part in improving the future of his country. Here we see the refugee portrayed as a disloyal and impotent deserter, unwilling or unable to defend the homeland when she needs help.[6]

3. Refugee as a focus of pity/stigma (Britain)

One (Bosnian) respondent to the pilot survey communicated to me that he found the term 'refugee', which I had used in the introduction to the original version of the survey, at best confusing and at worst offensive: 'I have been here for nearly fifteen years. I don't consider myself to be a refugee anymore and honestly I would prefer not to think again about that time when getting that letter [from Home Office] was such a problem for me.' In this example the stigma that can sometimes be associated with refugees was translated into a denial of association for the potential participant and a desire to 'move on' from that immigration status as a defining identity.

4. Refugee as a marker of strength and resilience (Britain)

In one interview with a Bosnian woman, Adela, who represented a Bosnian community organisation in London, I mentioned the use of the term 'refugee'. I described some of the reactions I had received to that term and asked what she thought about applying the term 'refugee' to herself. Adela responded:

> I feel really quite strongly about this. Why should I not call myself a refugee? That is what I am. I was happy there [in Bosnia], I didn't want

to leave. Pretending that I am not a refugee is like saying that never happened, that I didn't lose my home and that I wasn't forced to leave. I am proud of what I am. I have had to struggle and fight so hard for everything.

(Bosnian interviewee, female, age 40)

Adela has a strong identification with the term 'refugee' and considers her 'refugeehood' to be an illustration of her strength and resilience in the face of significant trauma. In the context of her employment assisting other Bosnian refugees, Adela has had a considerable degree of exposure to the suffering of her co-migrants which, combined with her own experience of trauma, has meant she has developed an acuteness to her understandable level of anger, an intensity which she has turned into embracing her status as a refugee almost as a matter of pride – a mark of defiance to her tormentors to prove that she has not been broken by an experience that she had neither wanted nor expected.

In the reported perception of local British residents through the lens of those who did consider themselves to be refugees we can see how both the contexts of departure and reception may have had had an effect on how some self-describe their status. In her memoir, Bosnian refugee Vesna Marić describes how on their departure from Mostar by bus to England, their Bosnian representative, Dragan, had gone through the group to make sure they looked enough like how he anticipated their British hosts expected a group of refugees to present themselves:

None of us knew exactly where we were going. All that Dragan had told us was to dress down and look as bedraggled as possible, because the previous group, he said, was too dressed up. The British had complained they didn't really look like refugees [...] The British had, understandably, expected something a little more like 'proper' refugees: people suffering, hardship visible on their faces, clothes torn and wrinkled, children's eyes crusted with tears.

(Marić, 2009: 28)

One survey respondent, a Serb woman from Bosnia who was 29 when she arrived in Britain in 1992 from Sarajevo articulated how she did not want to claim asylum at first:

When I arrived to the UK I was ashamed to claim refugee status and I think I was worried that I will never be able to go back if authorities in former Yugoslavia find out about it.

(Survey respondent, Bosnian Serb, female, 44)

In both the above narratives we can see how the migrant's ideas around refugeehood are juxtaposed against what they perceive to be expectations of, in the first narrative, those in the country of destination and in the second, worries related to perceptions in the country of origin. We can see through these articulations of refugee identity how the individual migrant's migration status and relationships towards the homeland and the new home are acting in mutually reinforcing and cyclical ways where the transnational space is both being dictated by the migration status and also going some way towards defining it.

The 'exile'

Certainly not all migrants from the former Yugoslavia and indeed far from the majority of the respondents in my research consider themselves to be a refugee, in any sense or interpretation of the term. As with Ana's migrant (lack of) identity above and Adela's strong affiliation with her 'refugee-ness', it is interesting to see how the identity of those from the former Yugoslavia in Britain vis-à-vis their position as a 'migrant' is reflected in the individual narratives and in particular related to the concept of living in exile.

Vesna Goldsworthy in her memoir *Chernobyl Strawberries*, which includes reflections on her move from Belgrade to Britain in the mid-1980s and her 'dual identities', considers some of the potential labels associated with having moved to another country and how they may apply to her situation. Goldsworthy, who moved to Britain in 1986 to marry an Englishman, is another whose experience of migration could essentially, like Ana's, be considered a 'love story':

> Unlike my ancestral matriarch and so many others in the part of the world I come from, I have never been a refugee. I am not an exile. Not quite an expatriate either: that term seems to be reserved for those coming from lands which are more fortunate than mine. A migrant, perhaps? That sounds too Mexican. An émigrée? Too Russian.
>
> (Goldsworthy, 2005: 36)

Said (2000) argues that the exile seeks to make sense and order out of his/her new world as a means of compensating for loss. Indeed, the term of 'exile' is so often associated in the contexts of Western Europe and the United States with the political or cultural output of those who have either 'chosen', or more usually, had that condition imposed upon them that it can carry a certain level of kudos, in a way that the term 'refugee' does not. Altenberg writing in Dumpor *et al.* (2005) states 'the words diaspora and exile carry negative connotations of scattering, break-ups, exclusion, displacement, seclusion and penalty', but I would suggest that the 'exile' concept, whilst undoubtedly carrying with it the association of loss, is acquiring in some circles almost a more fashionable status of a migrant who has opted for exclusion, often from repression, on the basis of his/her articulated ideologies. Said also refers to exile as a solitary position, the voice of a lone dissident compared with the mass of refugees in need: 'The word "refugee" has become a political one, suggesting large herds of innocent and bewildered people requiring urgent and international assistance, whereas "exile" carries with it, I think, a touch of solitude and spirituality' (Said, 2000: 144). We could take this further in the British context and conceptualise the perception of the exile as a donor – a producer or a creator of some kind of output, be it literary, artistic, political – whereas the refugee, in the popular discourse, is often caricatured as a beneficiary or a receiver of aid and assistance.[7]

Some migrant respondents have articulated how their choices around migration labelling have had an impact upon how they are perceived by others and on how they consider their position both relative to the host society and the context of departure from the homeland. One interviewee from Bosnia, on reflecting on the different labels that have been associated with his status in Britain offered the following thoughts:

> I would much prefer to be considered an exile than a refugee. I think the exile thing would give me more control somehow, like it's a choice, rather than a, 'shit we have to leave or we are going to be fucked' kind of thing, yeah, makes it sound more like an intellectual decision, one made over the freedom of a cup of coffee with a cigarette, instead of over a second bottle of rakija and a half-packed suitcase.
>
> (Bosnian interviewee, male, age 39)

Some aspects of the exile narrative are peppered through the accounts of those migrants who may not actually use the word 'exile' as such but whose experiences reflect in particular the tangible sense of loss so closely linked with the exile condition (Said, 2000).

Community absorption and 'diasporic borrowing'

Self-identification as a migrant/refugee/exile and also identification with the original region of origin also has implications for the ways in which members of the diaspora interact with each other, with non-migrants in Britain, with non-migrants in the homeland and with the British state. One Bosnian respondent, for example, fleeing the war in Bosnia in 1992, explained how he felt as though his original claim for asylum in Britain was rejected as at his asylum screening interview, in response to the question 'what is your country of origin', he replied 'Bosnia', a country which, despite the ongoing violent conflict, was a recognised state, whereas, some of his former compatriots fleeing non-conflict parts of the former Yugoslavia and on responding 'SFRY' were granted asylum on the grounds that there was no viable state to which they could be returned.

Croatian migrant Alberta offered some thoughts on her own migration experience in the context of that of some of her compatriots:

> The Home Office were dealing with asylum seekers and refugees – they wouldn't deal with the rest of us. I could have claimed asylum, plenty of others did but I refused to do that, I didn't want that on my passport – too much stigma. I knew a girl from Split, her family were Jewish and she claimed the family were being discriminated against so she could claim asylum. Some other guy from down the road claimed he was gay. I thought oh, what am I missing here?
>
> (Croatian interviewee, female, age 40)

It is evident from Alberta's narrative how the labels ascribed to themselves by some migrants can illustrate some 'tweaking' of status deemed necessary by some migrants to satisfy legalistic definitions of who and who is not to be granted leave to stay in Britain. We see evidence of some migrants who, in considering that their own individual circumstances, may not be sufficient to meet the criteria of the immigration authority's decision-making process, over who is to be granted entry and who is not. They sometimes 'borrow on' the collective experience and common narrative of their compatriots, even taking advantage of the confusion as Yugoslavia disintegrated, to equalise their experience with those in another country altogether. This can then cause understandable resentment on the part of those who have a full experience of the 'borrowed' migration story, possibly feeling as though their suffering has been hijacked by those whose political elites may have been party to or instigated that suffering in the first place. This type of borrowing and manipulation of the collective experience should be understood differently from another type of 'borrowed' experience where the diasporic narrative is absorbed sponge-like into the individual consciousness, especially prevalent in the articulations of those community representatives who spend considerable amounts of time listening to and sometimes advocating on behalf of their former compatriots. Choices and positioning around migration terminology can therefore play the role of challenger to the shared memory and collective experience deemed so crucial for the cohesion of the diasporic narrative.

Within the narratives of community representatives, it is possible to observe an accumulation of empathetic experience in the ways in which other employees and volunteers who, coming from the region, choose to either earn a living or volunteer to help support their (mainly) co-ethnic diasporic members in Britain. The move to peer support in working with those who have experienced trauma has been discussed in the literature in reference to refugees (Mestheneos, 2011). Such literature has mainly focussed on the ways in which those receiving assistance can benefit from having a support worker who could be considered to have shared some of their experiences (Moran *et al.*, 2011) and has also conceptualised such peer support as a tool to engender self-confidence and empowerment in the support worker as he/she comes to term with his/her own experiences. Less covered in the literature have been the potential effects on peer supporters in terms of how such repeated and regular exposure to the (often traumatic) experiences of others may merge with their own more personal histories.

Such regular contact with the personal narratives of other migrants was observed particularly through the experience of those respondents who, in their positioning of community representatives, often act as conduits through which other migrants articulate their hopes, ambitions, frustrations and disappointments related to the homeland, the new home, towards each other and towards other diasporas from the region. Through the narratives of some community representatives it was possible to observe ways in which their own personal histories were acting as a sponge absorbing the narratives of others with some aspects being borrowed in forming their own personal articulation of their relationship with their homeland and the new home. In this way, the transnational space of an individual can be visualised as being directly moulded by the shared or collective history diasporic experience. Whilst this was acutely articulated in the cases of those who regularly operate across the level of the community, it was also observed through the narratives of those migrants who may, at first glance, appear not to demonstrate any formal community ties but who may also 'borrow on' the collective narrative through other aspects of their socio-cultural choices – friendships, reading material, exposure to sport, music, art and so on.

Identifications and identity avoidance

The whole question of identity as related to transnational or diasporic migrant individuals and communities is one which has been the focus of much debate in the literature (Bash & Zezlina-Phillips, 2006; Bradatan *et al.*, 2010). To say that the question of identity relating to the former Yugoslavia is difficult to unpick would be a considerable understatement. It is important to hear the voices of those who have stressed the importance of not just focussing on issues around ethnic variables in the debates on transnationalism (Mitchell, 1997) and to note the warnings of those who caution against essentialising the 'migrant' aspect of identity over and above other elements that make up an individual's identity formation (Nyberg-Sorensen *et al.*, 2002a). Čolić-Peisker (2006: 220) for example conceptualises two different 'types' of transnationalism practised by Croatian migrants to Australia: 'ethnic transnationalism' where ethnicity is the 'centre of gravity' and 'cosmopolitan transnationalism' where the focal point of the migrant's life is the career and financial opportunities presented through the migrant's professional life.

In hearing the voices guarding against essentialism, it is also important to acknowledge that this was a project exploring transnational ties of those who have undergone different experiences of migration and whilst other aspects of identity were of course reflected through the narratives, these tended to be framed within the context of the migration experience – possibly because that was the focus of the discussion taking place. One exception to that would be articulations of the 'familial role' especially by women who emphasised their identities as mothers and grandmothers. The way in which the project was framed of course could have had the effect of excluding those migrants who considered themselves to not exhibit any ties to the country of origin, thereby essentially self-excluding from the project.

Much discussion has taken place in the literature around whether ethnicity and ethnic affiliation is self-determined or heritable (Cohen, 1999). Can it be considered a 'temporary moment of closure?'[8] Lucken (2010) in her thesis on identity formation of Bosnians in the US argues that 'Bosnians in exile reinterpret what it means to be Bosnian' and maintains that Bosnian national identity was not fully formed prior to migration.[9] There is however an inherent threat of misinterpretation in such statements. To what extent can such declarations about the nascent sense of Bosnian identity be hijacked by those, such as the British MP referred to in Chapter 2, who maintains that there is 'no such thing as a Bosnian'?

Respondent narratives reflect many articulations of 'international' identities, especially in those who have been living in Britain for a number of years and articulations of identity affiliation which could be conceptualised as something approaching 'a-national'.

> Do I feel Eastern European? Do I feel Slavonic? Do I feel former Yugoslav? Of course I do but I also feel more British than the British – I feel all those things at the same time.[10]
>
> I feel like [I] belong nowhere and belong everywhere. Could live anywhere. Don't feel like really belong here but definitely don't feel like belong there. No identity.
>
> (Survey respondent, Serbian, male, age 55)

Yugoslav affiliations

Nearly a quarter of the respondents to my survey stated that their country of origin was 'Yugoslavia' with many of those also stating their ethnicity as 'Yugoslav' and their nationality as 'Yugoslavian'. In exploring questions of ethnicity, nationality and country of origin with both survey respondents and interviewees, it was clear that there was still, amongst some participants, a strong affiliation with a Yugoslav(ian) identity. When I asked one (Bosnian) interviewee to reflect on why people living in Britain today, coming from the region may have identified so closely with Yugoslavia in their responses, he offered the following response: 'They are probably Serbs. Or from mixed marriages. Or Communists.'

Wachtel (1998: 2) draws on Anderson's (1983) *Imagined Communities* in viewing the Yugoslav nation 'not as a political entity but as a state of mind'. The terms 'Yugosphere' and 'Yugonostalgia' have entered the lexicon in recent years. The first, coined by Tim Judah in *The Economist* (although he has since expressed regrets over it), relates to 'an economic and social phenomenon' whereby the 'renewal of thousands of broken bonds across the former state' are gradually being enacted.[11] Judah is at pains to point out that Yugosphere is not to be confused with Yugonostalgia (Skrbiš, 1999). In his LSE paper on the Yugosphere (Judah, 2009), Judah focuses on the macro levels of economic, trade and organised crime elements of shared or common space across the area that was once Yugoslavia. In his analysis of micro-affiliations, issues affecting the daily lives of residents, Judah highlights how 'nation still trumps state' in the enactment of Balkan national spheres which cross state borders.

There was evidence of a similar understanding of ethnic spheres across parts of the region as expressed by my respondents. In conversation with one (Serbian) interviewee I mentioned another interviewee who was from Banja Luka, in the Republika Srpska part of Bosnia. His response was: 'Banja Luka? Isn't that Serbia these days?' However, those who fall most outside of Judah's Yugospherical vision of the Balkans are the former-Yugoslav Albanians, who he sees as forming part of an emerging 'dynamic Albanian sphere'. In Judah's analysis the Yugo spheres of influence do not reach as far as Kosovo (with the exception of Serb enclaves). Štiks (2006: 489) highlights how Kosovar Albanians were 'deprived of political and civil rights' under Serbian administrative rule and how difficult it was for some to even register as having Yugoslavian nationality.[12]

Given the strong Serb-Yugoslav link, it is perhaps not unsurprising to note that no respondents who had expressed any affiliation with Yugoslavia – either by nationality, ethnicity or country of origin – were from Kosovo. There is certainly an active (although possibly dwindling in recent years)

Yugoslav(ian) community within cyberspace reinforcing diasporic bonds across a number of host states. One way of interpreting such activity and identity could be associated with a rejection of recent nationalist past and its associated violence for some and yet of course, for those within the diaspora from Kosovo it is precisely the associations with Yugoslavia that are reminiscent of an oppressive, intolerant and nationalistic (under the Federal Republic of Yugoslavia) regime.

Identity avoidance

In his discourse on 'spoiled identity', Goffman (1963) includes the 'stigma of group identity'. In the case of some of my respondents, their identity tied with their ethnicity was expressed in terms of not conforming to ethnic stereotypes of what they perceive to be the archetypical personification of that ethnicity. I have already discussed how some of my participants expressed their doubts around contributing to the study as they felt as though they were not 'real' migrants and how identification with the term 'refugee' was avoided by others. Here we also see examples of respondents stating in the early stage of their interviews that they didn't feel as though they were 'true' representations of their ethnic group.

> I am not a typical Serb, I don't consciously look to hang out with the others, we have the priest round occasionally but we're not really religious or anything. You should go to the church in Ladbroke Grove – you'll get some rich pickings there.
>
> (Serbian interviewee, female, age 41)

> You want proper Serbs, I am not your man. You want the real thing – Ravna Gora[13] – that's where you want to go. If they'll speak to you – which they won't.[14]
>
> (Serbian interviewee, male, age 52)

In addition to the expressions on the part of migrants that they did not conform to ethnic stereotypes, some respondents referred to a battle against negative stereotypes, in reference to their ethnicity. Adrian from Priština stated:

> I see it in their faces. The reaction when you say you're from Kosovo. And you're Albanian. They think I either beat my wife, have her tied up in the back of some van somewhere or I carry drugs up my backside. Sorry, but you have touched a bit of a nerve there.
>
> (Kosovar interviewee, male, age 39)

There were a number of references by Serbs to perceptions around being unjustly labelled as the guilty party in all ethnic conflicts in the region and to the pressure they felt to take on personal responsibility for the acts of their nationalist political elites:[15]

> I met with quite a bit of hostility from my peers in school because I was Serbian and Serbians were seen (still are) as the bad guys of The War so I felt stigmatised by that. I had lots of people giving me abuse on that subject – 'Serbians are evil, it's monstrous what you do' – and I felt that Immigration Officials often treated me like scum and asked me invasive, unpleasant and irrelevant questions just because they could. I found it hard to be Serbian in the UK because I felt scapegoated, like I was in some small way personally responsible for the Bosnian War.
>
> (Survey respondent, Serbian, female, age 44)

> Well, we're all monstrous, murdering nationalists aren't we? I have lost count of the number of times that I have had to explain to people here that I don't support Milošević, I don't hate my neighbours and I am not about to murder anyone in their beds. I am also sick of having to tell people about the Serbs who have been killed and injured and lost their homes and all the people displaced and suffering. It is heartbreaking. But you never ever hear that side of things.
>
> (Serbian interviewee, female, age 37)

The levels of wariness illustrated in the following quotations go some way to justifying any reluctance to identify as a Serb – particularly amongst military-aged males.

> I don't trust anyone until I know exactly what they did and where they were [in the early 1990s]. If they were here then ok, fair enough but if they came after then what were they all doing? They can't all have been army cooks.
>
> (Bosnian interviewee, male, age 40)

A similar comment about the number of 'chefs' in the Bosnian war was made by Danis during his interview:

> Look, am not saying for a minute that I don't have Serb friends, of course, I do. Some of my best friends are Serbian [laughs]. But I need some proof you know, about their background I mean. If you believe every Serb you speak to in London, that fucking army must have spent the whole war either eating or cooking.
>
> (Bosnian interviewee, male, age 37)

Partly in response to reactions as described above, an explicit identity avoidance tactic was expressed by some in terms of 'morphing' their ethnicity into another in a situation where they predicted a hostile reaction. One Serb interviewee told me that when she first arrived in Britain she would tell people she was Russian to avoid having to answer questions about the war in Yugoslavia. Goldsworthy (2005: 209) also narrates the story of telling those who asked that she was from Russia.

Conversely, others when confronted with situations where they are likely to face hostility related to their ethnic identity take strength in the group identity (similar to those feelings expressed by Serbs in response to the NATO bombardment of Serbia). Serbian, Jana, who describes herself as having an 'international' identity spoke about her experience as a member of the diaspora in Britain during the time of Yugoslavia and sheds some light on how the diasporic dynamic changed as Yugoslavia disintegrated.

> I think it was 1994 or 1995 there was a group of Bosnians doing a jumble sale. I spent days trying to decide whether to go or not, I collected some stuff and off I went and the look on those women's faces when they saw me […] I didn't stay and I cried in the car on the way home. And then another time, I also went to the big cathedral to a Croatian service at Christmas. The priest, he was using the Bible to point out that Serbs were the worst nation on the planet. I couldn't listen to it and I left feeling absolutely terrible. Afterwards I met the Serbian priest at the Embassy who had also been there that night and I talked to him about it. He said to me that if he had known that I was there, he would have felt stronger.
>
> (Serbian interviewee, female, age 60)

The reference made by Jana to the priest's comments of 'he would have felt stronger' reflects the solace and comfort that those who feel victimised can take through the strength of diasporic bonds.

Narrating guilt and the migration experience

Guilt as articulated by migrants from the former Yugoslavia can be conceptualised in two ways. The first is as an expression of transnationalism itself in the ways in which, within their feelings of guilt, are contained a complex and multi-layered multitude of connections and relationships with families, friends and with the 'home' in general. The emotion of guilt is thereby an emotional 'product' – an outcome of the migration experience often in unexpected ways. My second understanding of the different articulations of guilt is as a conduit through which other transnational acts are executed. Guilt often took on the role of the assertive sponsor behind much of the (especially) economic transnational activity being expressed by migrants.

Vesna Goldworthy, in discussing her parents building a large house in Serbia which she and her sister were expected to fill with grandchildren, describes their 'failure' to deliver on this dream in the context of the guilt she felt:

> Whenever I think about this, it hurts. I've never even thought, not for a second, of fulfilling their longing for me to go forth – or rather, to stay put – and multiply. I entertained the very Western idea that my first responsibility is towards my own happiness. I have been paying for this presumption with the small change of guilt every now and then. Of all the languages I know, guilt is the one that my memory speaks most fluently.
>
> (Goldsworthy, 2005: 7)

This feeling of guilt situated within the parent–child relationship was one that resonated throughout the interviews I conducted irrespective of immigration status, age, gender, country of origin or time of arrival in Britain. It is usually expressed within the context of guilt around departure, as evidenced by the following quote from Bosnian refugee Nedim in describing his decision to leave and the ongoing dynamic of his relationship with his parents:

> I didn't feel as though I had much choice to leave to be honest. I could see the way it was going over there. I saw people coming back [from fighting]. I could see what it was doing to them. My parents were in their '60s then and they really wanted me to go, there was absolutely no pressure to stay, the exact opposite in fact. And for my brother too. So, I came to London and got on ok more or less. But, at the back of mind was always, what kind of shit leaves his mother in what was back there. I couldn't shake it. It was starting to affect everything, I couldn't have a girlfriend, there was no way I was going to make her deal with all my shit. And then they got to Croatia and it was a bit better, I didn't have the immediate, are they going to live or die worry at night. But what kept me awake was the thought that I didn't do enough, I didn't do enough to get them to England. I was a mess. And now, in some ways it's even worse. My parents are old. We can't even talk on the phone much 'cos Dad's going a bit … you know … so now I am thinking it will be better once they've gone. And what happens then? More

guilt. So basically I am fucked either way. And I haven't even managed to give them any grandkids because I am so messed up. At least my brother's managed that.

(Bosnian interviewee, male, age 36)

Nedim articulates very clearly what other respondents also expressed – the guilt at leaving family members behind with the uncertain knowledge of what might face them. Even when those family members were encouraging of their departure and even strongly supported it, those who had made the journey to physical safety and had, for different reasons, left behind close family members were dealing with sometimes intense feelings of guilt even 20 years later and long after the physical threat had passed.

Guilt around departure associated with the familial relationship was a consistent thread through several interviews. Alma, also from Bosnia, described the role which guilt, stirred by parental insistence, played in her migration decisions and in her interaction with her friends:

If it was up to me, I would have just stayed. That's part of the whole guilt thing. Now when I talk about the war with friends who stayed you can see this shadow over their eyes. I know there's things they're not telling me but that's exactly what I mean when I say you can't really know, and don't listen to anyone from here – the diaspora I mean – who claim to tell you how it was. Just like over there they don't know what it's like to leave, we don't know how it was for them. Not really. Not as a daily experience. So I feel guilty about that. Because we went from sharing everything, all the usual teenage really intense stuff to now, we are just not on the same planet, do you know what I mean? We have the common background but we can't get past those shadows. And I have shadows of my own. Maybe not as long as theirs but they're still shadows. And I wouldn't have even left but I couldn't stand what that was doing to my parents. I could see them getting more and more eaten up by the whole thing. With the worry I mean. And it would have absolutely destroyed them if something happened to me, I know that. So I left and now I have guilt from the other side. You can't win.

(Bosnian interviewee, female, age 37)

Alma's expressed guilt is also multi-layered in that she attributes her original decision to leave her home out of guilt at her parents' distress and anxiety at her refusal to leave and then subsequent feelings of guilt at not staying and 'seeing it through' with her friends. Such expressions of guilt related to the homeland and its associated 'abandonment' have formed webs in the lives of some migrants, which are placed across other aspects of day-to-day existence in Britain, particularly relationships with friends and family members – both 'over there' and 'over here'.

Economic expressions of transnationalism motivated by guilt were also expressed by Vanja from Belgrade in terms of buying expensive gifts for friends and family 'left behind':

> Yes, I go back all the time but it is not the same. I am always bringing stuff because I feel guilty. Ah, the guilt. Nobody tells you about that when they are putting you on the bus. But then how would they know, it's not as though any of us had been through something like that before.
>
> (Serbian interviewee, female, age 35)

Ana attributes guilt as the main motivation behind accepting a job in a Yugoslav company in London. Ana felt as though she 'owed' something back to the authorities who had paid for her language training in London and whom she felt as though she had somehow 'betrayed' by running away to get married to an Englishman. Ana eventually worked for that company in different forms for a period of 20 years and felt when she retired that she had 'repaid her debt'. We can also see perhaps shades of Ana's feelings of guilt in her eagerness to come to the assistance of those migrants coming from Serbia after her: guilt perhaps because she left at a different time when things were not so economically difficult in her home country and others coming after her were subject to the privations of sanctions and NATO bombing.

Others have articulated how guilt at the action of others has contributed to their individual sense of identity related to the country of origin.

> My news bulletins at the BBC were exemplary. Outwardly, I kept my distance and knew how to be even-handed. In my feelings about the war, however, I tended to overcompensate both in my distress that Serb suffering did not seem to register anywhere and in my shame that the Serbs could cause so much pain to others. Both suggested that my relationship with my own Serbianness was perhaps more raw than I admitted even to myself. It was part of a knotted circle of love and guilt which I preferred not to pick at very much.
>
> (Goldsworthy, 2005: 213)

The guilt narrative is one which is complicated to unravel and permeates throughout aspects of the life of the migrant but especially in the ways in which he/she develops a relationship with the homeland and those associated with it.

Whilst it is understandable that those working within the migration sector – be they policy makers, academics, journalists or parliamentarians – seek to understand and conceptualise the migration experience (and that is after all what this project was partly based on), the voices of those who migrate – with whatever motivation(s) – can often cut across all the debates in the abstract. The experiences of those from the former Yugoslavia have demonstrated how personal identifications related to the status of being a 'migrant' may not always tally with any classification by others – as well-intended as such classification may be.

Notes

1   A. Chrisafis, 'At night it's like a horror movie', *Guardian*, 6 April 2015.
2   For examples see the Migration Observatory (University of Oxford) 2013 report 'Migration in the News' and a series of online essays, also the 'New Keywords Project' coordinated by Nicholas De Genova and Martina Tazzioli at King's College London: http://nearfuturesonline.org/europecrisis-new-keywords-of-crisis-in-and-of-europe.
3   Although I recognise that in juxtaposing the term of 'migrant' with 'economic', 'labour' or 'career', the media discourse on migration is now creeping into more academic usage with its distinction between 'refugees' and 'migrants'. I am however using the term of 'migrant' solely to describe the status of an individual who is living in a country other than the one in which he/she were born.
4   www.geiriadur.net/index.php?page=ateb&term=Hiraeth&direction=we&type=all&whichpart=exact (accessed 27 July 2014).
5   In their attempts to reduce net migration, the British government's express intention through the introduction of the 2014 Immigration Act was to create a 'hostile environment' for new migrants.
6   This kind of caricaturing has resonances with some of the British tabloid and other popular discourse around the increasing numbers of forced migrants across Europe, insinuating from those pictures in the media showing large numbers of male migrants, that women and children are being 'abandoned' in war zones.
7   There have been a number of campaigns by NGOs targeted at dispelling misconceptions of refugees and migration in general.
8   Ethnic diversity in social science and public policy research workshop, London Metropolitan University, 27 April 2010.
9   Kirsten Lucken seminar at the LSE, 'Bosnian Muslims in New England: Identity maintenance and integration patterns', Forum on Religion Seminars, 5 May 2010.
10   Interview recorded for the Evelyn Oldfield Unit Refugee Communities History Project and archived at the Museum of London (reference 2005.120).
11   Tim Judah, 'Let's hear it for the Yugosphere', *The Economist*, 23 June 2011: www.economist.com/blogs/easternapproaches/2011/06/former-yugoslavia (accessed 22 April 2016).
12   Given the lack of identification and affiliation on the part of Kosovar migrants with Yugoslavia and the way in which the call for participants in this project was framed, it is, in a way, not unexpected that the study did not generate more interest in potential respondents from Kosovo.

13 Hotel in west London named after an area in Serbia known for its association with the chetnik movement.
14 They didn't.
15 The *Britić* online archives for example show that the magazine has also been involved in another initiative, 'The Serb Lobby', which asks why are Serbs always the 'bad guys' and states as its aim: 'fairness and accuracy in politics and the media. Let's rebrand Serbs! We're the good guys'.

References

Anderson, B. (1983) *Imagined Communities: Reflections on the Origins and Spread of Nationalism*. London: Verso.
Bash, L. and Zezlina-Phillips, E. (2006) 'Identity, boundary and schooling: Perspectives on the experiences and perceptions of refugee children', *Intercultural Education* 17(1): 113–128.
Berg, M. (2011) *Diasporic Generations: Memory, Politics and Nation among Cubans in Spain*. Oxford: Berghahn Books.
Bradatan, C., Melton, R. and Popan, A. (2010) 'Transnationality as a fluid social identity', *Social Identities* 16(2): 169–178.
Cohen, R. (1999) 'The making of ethnicity: A modest defence of primordialism', in E. Mortimer and R. Fine (eds) *People, Nation and State*. London: I.B. Tauris.
Čolić-Peisker, V. (2006) ' "Ethnic" and "Cosmopolitan" transnationalism: Two cohorts of Croatian immigrants in Australia', *Migracijske i etničke teme* 22(3): 211–230.
Dumpor, A., Mahmutović, A., Mešković, A., Tojčić, I. and Altenberg, K. (2005) 'The art of identity: Memory, myth and a feeling of home'. London: British Council.
Goffman, E. (1963) *Stigma: Notes on the Management of Spoiled Identity*. New York: Touchstone.
Goldsworthy, V. (2005) *Chernobyl Strawberries: A Memoir*. London: Atlantic Books.
Halilovich, H. (2013) *Places of Pain. Forced Displacement, Popular Memory and Trans-Local Identities in Bosnian War-Torn Communities*. New York: Berghahn.
Judah, T. (2009) 'Yugoslavia is dead: Long live the Yugosphere'. LSEE Papers.
Levitt, P. and Glick Schiller, N. (2004) 'Conceptualizing simultaneity: A transnational social field perspective on society', *International Migration Review* 38(3): 1002–1039.
Lucken, K. (2010) 'Identity matters: Bosnian identity maintenance in a post-migration setting'. PhD thesis, Boston University.
Marić, V. (2009) *Bluebird: A Memoir*. London: Granta.
Mestheneos, E. (2011) 'Refugees as researchers: Experiences from the project "Bridges and fences: paths to refugee integration in the EU"', in B. Temple and R. Moran (eds) *Doing Research with Refugees*. Bristol: Policy Press.
Mitchell, K. (1997) 'Transnational discourse: Bringing geography back in', *Antipode* 29(2): 101–114.
Moran, R., Mohamed, Z. and Lovel, H. (2011) 'Breaking the silence: Participatory research processes about health with Somali refugee people seeking asylum', in

B. Temple and R. Moran (eds) (2011) *Doing Research with Refugees*. Bristol: Policy Press.

Nyberg-Sorensen, N., Van Hear, N. and Engberg-Pedersen, P. (2002a) 'The migration–development nexus: Evidence and policy options', *International Migration* 40(5): 49–73.

Richmond, A. (1994) *Global Apartheid*. Oxford: Oxford University Press.

Said, E. (2000) *Reflections on Exile and Other Literary and Cultural Essays*. London: Granta.

Skrbiš, Z. (1999) *Long-Distance Nationalism: Diasporas, Homelands and Identities*. Hants: Ashgate.

Štiks, I. (2006) 'Nationality and citizenship in the Former Yugoslavia: From disintegration to European integration', *Southeast European and Black Sea Studies* 6(4): 483–500.

Turton, D. (2003) 'Conceptualising forced migration', RSC working paper no. 12, Queen Elizabeth House, University of Oxford.

Van Hear, N. (1998) *New Diasporas: The Mass Exodus, Dispersal and Regrouping of Migrant Communities*. London: University College London Press.

Wachtel, A. (1998) *Making a Nation, Breaking a Nation. Literature and Cultural Politics in Yugoslavia*. Stanford, CA: Stanford University Press.

4   Intangible transnationalisms
The allegory of dreams

For the sake of a heaven's realm
The common life has been propelled into the realm of dreams
Some victims still believe it's just another nightmare
But Saharan borders are being drawn for real
In a country full of water and wood –
Overnight lifeless, overcrowded only by ghosts.

Jančić (1996: 8)

The allegory of the dream is one which runs through much of the discourse on migration, permeating both academic and more popular narratives (Glick Schiller & Fouron, 2002). Visual and printed media on the experiences of those leaving their countries of origin in search of a new life elsewhere make emotive references to the dreams of migrants, often in the context of 'broken' dreams when things do not go as planned, harrowing descriptions of journeys to achieve dreams or stories of 'dreams' versus 'reality'. Academic literature on migration is also replete with references to the dreams of migrants as motivators or catalysts for making the move to a new country (Wang, 2013: van Meeteren, 2012; Parutis, 2011). As in the media, the focus in the academic and more popular literature is also often on unrealised dreams, the fracture between imaginings of a better life elsewhere and the reality of the migration experience (Du Phuoc & Ricard, 1996). However, the dreams of migrants do not always refer to the waking imaginings of future lives in the metaphorical, aspirational sense. Some migrants' dreams are not coloured by career ambitions, house building plans or blossoming familial lives. For some it is the echoes of past lives and those lost which permeate their subconscious. Such dreams are made more vivid by the sad reality that these are not wholly imagined lives; they are lives that once were and imaginings of lives which could have been had their trajectories not been knocked off course by the experience of (forced) migration.

This chapter therefore first presents some context of the ways in which the dream narrative features within the academic and more popular discourse on migration before discussing how some 'dream-related' features have coloured the presentation of the experiences of those fleeing the conflicts of the former Yugoslavia, particularly in the field of mental health. The dream narrative was very strong in the narrated experiences of my respondents. I discuss some examples before considering how the juxtaposition of social and mental spaces has already been treated, particularly in relation to refugees from Bosnia and how these may impact upon the formation of a transnational dynamic.

The representation of the dream allegory in the migration narrative

Within the popular discourse on migration, the country of destination is often portrayed as the land where dreams can come true, the process of migration is characterised as an aspirational journey in a metaphorical sense and the migrant him/herself is considered to be in pursuit of some kind of dream – usually in search of a 'better life' (often in a financial sense but could also be aspirations related to education, career, safety, security, health or improved family life). As immigration legislation across Europe becomes ever more restrictive with those who wish to seek refuge being forced into increasingly desperate means of gaining access to a safe haven, the media focus has been on the fracture between the idealised dream of those seeking refuge, compared with the stark reality. There is a plethora of headlines which draw on the allegory of hopes and dreams of migrants and crucially, the abandonment of such dreams in the face of increasingly draconian measures taken by states to inhibit the realisation of any such aspirations.[1] In recognition of the gulf between desire and reality, the dream allegory is then transposed into a tale of nightmares.

There is also a thread of references to dreams and dreaming running through the migration discourse within the academic field with some observers similarly focussing on the 'dream versus nightmare' dichotomy (Christou, 2006). The theme of (broken) dreams related to the migration experience seems to focus in on three particular groups of migrants: international football players (Esson, 2013; Poli, 2006), sex workers (Chimienti, 2010, Levy & Lieber, 2008) and those trafficked into sexual exploitation (Derluyn and Broekaert, 2005). The experiences of those who dream of a new life far from home have also been the subject of debate (Bal & Willems, 2014; Horst, 2006). Some migrants seem to have accepted that their own personal dreams are not to be realised but that the act of migration has put them in a position to help the next generation achieve their goals (Cartwright,

2011). The references discussed above are mainly focussed on the dreams and hopes of migrants in the aspirational sense. Allusions to a better life elsewhere and even the portrayal of broken dreams, disappointment and dashed hopes still refer to those would-be dreams or dreams that actual migrants are making in relation to their future lives in a new country of origin. There are also however some more limited examples of the ambiguity which can colour the dreams of migrants: the complexity of aspirations for the future combined with yearnings for the past has been the focus especially in the context of Chinese migration to the United States (Teo, 2011).

The mental health focus of the refugee experience

The experiences of those from the former Yugoslavia have been the focus of clinicians and health care professionals who seek to understand the effects of exposure to violence on the short and long term mental health particularly of those fleeing the conflicts in Bosnia and Kosovo. Such clinical studies have mainly focussed on prevalence of post-traumatic stress disorder (PTSD) amongst migrating populations (Priebe *et al.*, 2009; Weine *et al.*, 1998). Other studies have also investigated symptomatology associated with prolonged grief disorder (Morina *et al.*, 2010), adult separation anxiety disorder (Silove *et al.*, 2010) and other mental health disorders such as anxiety and depression. Reported prevalence rates vary significantly depending on the method of assessment with measurements of PTSD and depression differing significantly when the assessment has been carried out via a self-report checklist as opposed to a diagnostic interview (Mollica *et al.*, 2013).

Many studies find a positive association between gender and age and the prevalence of certain mental health disorders. Bogić *et al.* (2012: 218) for example find that higher rates of mood disorders were associated with the female gender, and older age and higher rates of PTSD were associated with older age. Weine *et al.* (1998) found an association between older age and rates and severity of PTSD diagnoses at baseline assessment. Vojvoda *et al.* (2008) in their study of PTSD in Bosnian refugees in the US identify women and older persons as being in particular need of 'appropriate and timely treatment' and Eytan *et al.* (2004) also find a positive association between female gender and older age with 'post-conflict symptoms'. Priebe *et al.* (2012) found older age to be associated with 'higher levels of both general psychological and posttraumatic stress symptoms' (Priebe *et al.*, 2012: 45) and that 'the odds of experiencing a higher level of PTSD symptoms were significantly greater in females, older persons and those who had experienced more traumatic events before the war' (ibid.: 52). They also reported that 'older age, female gender, not being married or co-habiting and being without employment were all associated with increased symptom severity' (ibid.: 53).

However, in asserting such a relationship between the presence of certain mental health disorders and gender, there appears to be little acknowledgement of the gendered difference in willingness to disclose symptoms associated with mental (ill) health. Such potential differences could be key contributing factors especially in studies where the assessment relies on self-reporting methods. Johnson and Thompson (2008) in their review of PTSD in civilian adult survivors of war trauma highlight the difference in PTSD prevalence rates in studies that have employed a self-reporting measure (which tend to show higher prevalence rates) compared with those who have been assessed via psychiatric interview. In the different (but also migration-related) field of trafficking, the gendered difference in clients of trafficking support services in terms of discussing mental health support needs and requesting the assistance of counselling and other mental health specialist services, is becoming increasingly recognised (Munro & Pritchard, 2013). In the reported, anecdotal experience of support providers for trafficking victims, it is not that men are less likely to experience trauma-related mental health pathology but that they are less likely to disclose any symptoms to a mental health professional or a support provider and any disclosure is likely to take a longer period of time.

However, it is not only the pre-migration experience of trauma which needs to be taken into account. In much of the literature discussing the mental health of refugees, forced migrants or those fleeing violence, there is an increasing acknowledgment of the role which post-migration stressors can play in the development of mental health disorders. Factors such as unemployment or denial of right to employment, financial difficulties, insecure or impermanent migration status (and the associated threat of repatriation), weak social networks and/or less well-developed language skills in the host environment have all been identified as potentially contributing to poorer mental health outcomes (Teodorescu *et al.*, 2012; Bogić *et al.*, 2012; Johnson & Thompson, 2008; Fozdar, 2009).

Dreaming of 'home': The juxtaposition of 'there/then' and 'here/now'

The relationship between sleep disruption and mental health, and in particular PTSD, has been well established (Pillar *et al.*, 2000). Although usually referring explicitly to the experiences of refugees, a number of studies have also looked at predictors of PTSD within non-asylum seeking migrant populations. A comparative study (Knipscheer *et al.*, 2009) by clinical psychologists in the Netherlands of post-traumatic stress symptoms as demonstrated by 'economic' or 'labour' migrants and 'political' migrants (as termed in their research) concluded, from the high levels of trauma-exposed economic migrants, that service providers and policy makers should not assume that those migrants deemed to have migrated 'voluntarily' would not also be presenting with symptoms of post-traumatic stress. This pathologisation of the migrant experience is not always to be welcomed. Sleep disturbance in itself, especially in the short-term, is not an unexpected reaction to the experience of migration. Considering the number of insomniac tourists in hotels around the globe, disturbed sleep could be considered a 'normal' and temporary reaction to the experience of displacement (forced or otherwise). But what happens when disturbed sleep patterns replace 'the norm' or become more prolonged, sometimes for years following the original migration journey? Reflected in the reported experiences of migrants from the former Yugoslavia were not only the stories of night-time dreams related to the former home, to family members killed or 'lost' to far-flung countries and to the experience of departure, but also the narratives of day-time (semi-) awake remembering of lives and family. The effects of such 'day dreams' can sometimes be extremely unpleasant when the dreamer not so much 'awakes' but 'comes to' reality and experiences the jarring effect of life in the dream and life in the day-to-day real world.

Ana talks about how an experience in her childhood is a feature of her dreams even now, sixty years later:

> The Prime Minister at that time was an anti-communist and he gave up some cells in exchange for Serbs and Slovenes in Italian camps. We were there as a family for a while when I was very young. My father was beaten and tortured in the camp. I used to have a dream later on [as an older child] about somebody opening and closing curtains. I had that dream lots of nights. I told my mother about it, she said I was too young to remember but at the camp they put me and my mother and sister on one side of the room and my father on the other with the curtains drawn. They would beat him then, open the curtains so we could see him then close them again and continue the beating so we could hear but not see. I still have that dream sometimes when I am feeling low.
>
> (Serbian interviewee, female, age 68)

In her memoir *Bluebird*, Vesna Marić, who came to Britain in 1992 as a sixteen-year-old from Mostar, places the following passage of her reflection on her dreams as a chapter of its own under the title 'Shadows':

> I had nightmares. One was persistent: empty streets and the clicking of heels. I was hiding, the war was raging and I had nowhere to go. England was only a dream. There was a man, a silver pistol at his hip, a slender man in white tennis shoes, blue jeans and a brown T-shirt.

I'd never seen him before. He'd say: 'I am the highest authority', and I'd wake up, sweat burning my face. I'd look at the wall, and remember I was in England, that I was a refugee. Relief.

(Marić, 2009: 67)

In this narrative we can see a double allegory of dreams of the homeland and desired home: the dreamer in her dream is picturing the new home as something to be aspired towards, something to be 'dreamt' of. Many respondents talked about dreaming that they were back in the homeland which, within those dreams, represented places of fear and aggression, spaces to be escaped from, and then articulated their sense of relief at waking to discover that they were in a different environment. The writer of the extract included above and those who articulated similar feelings of relief at waking from nightmares into a space of 'safety' tended to be from those who migrated at a young age. The narratives of older refugees from Bosnia who have made their homes – reluctantly – in Britain reflect a different kind of dreaming which, in some cases, threaten to overwhelm their day-to-day lives. On one level it is possible to see references to dreams related to the homeland in the cultural output of migrants from the former Yugoslavia who have moved to Britain and who use their creativity as an output for reflecting their subconscious images of the homeland. On another, possibly more poignant level, we can see how the dreams of the homeland are almost an intrusion on the everyday reality of life in the new home, to the extent that those dreams, in a way, *become* the everyday reality and the individual migrant becomes almost trapped in a state of extreme transnationalism – physically present in one location, within the borders of one territorial state and yet emotionally and mentally in another place and time entirely.

Adina who was 55 when she came to London in 1993 from Sarajevo articulates how her night-time subconscious 'back home' is affecting her daily life in London.

Every time I go back, it's like am reliving it all over again, the whole nightmare. Sometimes I think I should just stay here but I can't do that, not going back would make me so desperate. I am always there in my head anyway. I haven't even changed my watch – look! [...] Sometimes, in the morning when I wake up, in that split second before I open my eyes I think I am over there still, at home, with my husband making coffee, my son getting ready for work and my daughter fussing with her hair. Then, it's all gone. And I remember it all again. My husband is dead. My son is on the other side of the world and I have grandchildren who I see twice a year and who speak a different language from me. It's not easy for me, I was older when the war came.

It's not so easy to start all over again. To learn everything again. So, I go back whenever I can and it's ok when am there and then I come back and it's like missing a limb again. All over again.

(Bosnian interviewee, female, age 68)

Adina's feelings on 'waking' from her dream are far from the relief felt by others for whom the subconscious homeland represents a space of fear. For Adina the imaginary homeland is rooted within her very real and concrete memories of the shards of her once very full family life. Adina may well consider her homeland to be a site to escape from but that is not what permeates her dreams. Adina's reference to not changing her watch from Central European time has resonances with the respondents in the study by Bajić-Hajduković (2008) of Belgrade parents and their migrant children where those parents left behind in Belgrade referred to keeping one clock in the house set to whichever time zone their child is living in to help them feel somehow 'closer' and less physically and emotionally distant. In the case of those parents in Bajić-Hajduković's study it was the non-migrants who were aligning their lives to those of the migrants. In Adina's case she is the migrant who feels at every return to life in Britain that she is being dragged back into a world where she doesn't want to be. Reflections on missed opportunities are partly what intrude into Adina's daily life in London – not the missed opportunity of professional and personal development which may characterise the (forced) migration experience of younger migrants but through the narratives of Adina and others we see glimpses of imagined lives.

A number of respondents made reference to the age of migrants as being key to apparently successful outcomes in terms of 'integration', with a focus on the ability of younger migrants to learn the language, maximise education and training opportunities and just generally 'get on with life'. However, the factor of age alone doesn't explain the apparent difference in outcomes of integration without exploring some of the other dynamics at play. The structure and expectations of family life may have meant that some, in being forced to flee their homes, could see their family life unravelling before their eyes. Parents who were in or approaching retirement at the time of migration, possibly already grandparents or looking forward to grandchildren, would have been envisaging a totally different kind of life than the one they are living now. Through those imagined lives – some of which are reflected in the act of dreams – we can see how memories and reimaginings of life 'over there' can intrude upon life 'over here' as constant reminders of a life lost, hinting at another life that could have been. There were constant references through the narratives to how life 'might have been' if circumstances had been different. Some references are stronger than others and often involve family members – those who have died, those who live far away or, in some cases, those not yet born.

> I do think all the time about what I would be doing now if things had been different. I don't belong here but I have absolutely nothing to go back for. It's not good over there any more, it's all changed.
>
> (Survey respondent, Bosnian, female, age 44)

> All the family are scattered now. There are going to be [family name] all over the world! I am trying to get on with my life but it's hard you know. It would all be so different if we lived closer, I feel as though we are trying to fit a whole year of happy memories into a 2-week visit. Even the normal family arguments I miss. We are so polite to each other now, that's not what family is like. If we were still over there all of us, I could say whatever I want, now I keep my mouth shut because I don't want to spoil the holiday. We are like acquaintances, like work-mates, not proper family at all.
>
> (Croatian interviewee, female, age 38)

> Our family is shattered. I have three children and I never see any of them. You have no idea what that has done to me, how that has changed me. I want to be the grandmother they all come to after school for snacks and love. I'm not even a grandmother yet anyway because my children are all too busy building their careers. But what happens when they do have kids? Am never going to see them. My kitchen should be full of smells of the family life. And the only thing I can smell in my kitchen here is my own cigarettes and the curry from next door.
>
> (Bosnian interviewee, female, age 57)

Through the three narratives above we can hear the echo of the 'other' lives behind the realities of the lived experience, blueprints of which have left clear marks on the present everyday life of those who had imagined and still imagine other realities. In circumstances such as these it is difficult to conceptualise the transnational experience as anything other than spaces of at best, resignation and reluctant acceptance but at worst, of sadness, confusion, bewilderment and bitterness: a world away from the empowered career professional or entrepreneur. We can also see ways in which the family dynamics have been fundamentally altered beyond recognition with again the reality representing something far removed from the expected.

Whilst the bulk of the literature concerning gender politics in southeast Europe will couch any analysis of the gender dynamic as being a patriarchal discourse within the wider societal context, when it comes to the microcosm of the household space, it is the matriarch – specifically the Grandmother – who commands the respect of the family unit and who essentially acts as a figure of control – especially over any children and younger women within the household. Simić (1999) within the context of his research on gendered dynamics within the 'traditional Yugoslav family', sets out the importance of the place occupied by the oldest female in the household. Those women who in the early 1990s were at the point of the lifecycle where they may have been looking forward to 'promotion' into the place of household matriarch, have found that in addition to everything else they have had to cope with – including in some cases the death of family members, the loss of their home and the forced necessity to make a new life for themselves in a country where they may not even have spoken the language – have an additional dimension to their pain. For these women what has been taken from them has been not only the home, the community, the family – what has also been taken from them is their status in the family, their anticipated 'turn' as 'alpha' female.

Returning to Adina's story then, she made a number of references in her narrative to wanting to work and 'get on with her life' but clearly perceived her mental and emotional health to be a barrier to 'moving on' in such a way. Another interviewee, Adrijana, also talked about how dreams of old life interrupt and intrude upon her life in the present:

> I don't have a job anymore so sometimes I kind of doze off at home especially after lunch. I try not because the way I feel when I wake up is absolutely horrible. When I am asleep, it's all ok, am there, with everybody and it's all ok. And then I wake up. And it's not ok. I am alone and there's just no point to anything anymore. And what can I do then? Nothing, it's all a misery, I wouldn't wish it on my worst enemy.
>
> (Adrijana, Croatian Serb interviewee, female, age 55)

For both Adina and Adrijana, their dreams and memories represent a temporary refuge. The shattering of such a refuge however on waking is a cruelty which they continue to live with more than twenty years later.

Some of these articulations of transnationalism are indicative of how the duality of the transnational experience can sometimes become overwhelming. Summerfield (2003: 264) in his clinical work with Bosnian refugees in London articulated what he has observed as 'a refusal to accommodate to a world which now seemed unintelligible', which resonates with the following:

> I come and sit by the sewing machine and if I am not looking where I am, I am there, in my thoughts I am there. But when I look around

only then do I realise that I am here. I've never, ever, it's now 10 years since we arrived. I've never had a dream that was set here. I haven't dreamt of it, I either dream that I am in my neighbourhood or there where I was born, basically never anything set here. It's as if I never saw any of this, it never enters my dreams.[2]

We can see through these narratives how extreme transnational echoes of past and imagined lives can lead to distress and confusion in present day-to-day lived 'reality'. Far from enabling active and positive contributions to the societies of two nation-states, at these moments the transnational space can evoke conditions of debilitation and destruction which can affect the mental wellbeing of the migrant inhabiting that space. A number of my respondents spoke clearly about their motivations to move to Britain to further career aspirations or 'simply' to experience life in another country. However, it was the contrast between the dreams of those who appear to be living their lives out within, not so much the past but in a fictional present, compared with the way in which the 'hopes and dreams' narrative is presented within the popular discourse on migration which was particularly striking.

Building on the need to acknowledge the post-migration environment as a key variable in determining mental health outcomes, several voices have also warned against the dangers of over-pathologising the experiences of forced migrants and in particular those fleeing the conflicts of the former Yugoslavia. Halilovich (2013) in his work with Bosnian refugees rejects the stigmatisation that labels such as 'trauma' and PTSD can bring (Halilovich, 2013: 13). He views the 'invisible injuries of soul' experienced by his informants not as 'psychopathological and clinical conditions that needed to be or could be corrected' but through the act of telling their stories he sees 'normal, human responses – coping mechanisms – to the extraordinary situations and experiences these people had gone through' (ibid.). Fozdar argues that her respondents 'see much of the negative affect associated with settlement as a "normal" response to a difficult situation, rather than a pathological response in need of psychiatric intervention' (Fozdar, 2009: 1338). Fozdar concludes by postulating that 'by treating the emotional distress produced by their situation as "depression", one risks treating the symptom, rather than the cause, and pathologising an entire population' (ibid.: 1349).

Summerfield emphasises the *social* over the *medical* nature of the distress experienced by refugees from Bosnia, especially regarding recovery in the post-war period. He highlights the differences between what he conceptualises as 'medicotherapeutic' and 'sociomoral' ways of understanding the experiences of those who fled the conflict and who are still dealing with the aftermath of a lack of justice and a denial of responsibility on the part of the perpetrators. Talk of reconciliation rings hollow within a post-conflict environment where so many of the perpetrators of war crimes are not only free from impunity but are still occupying or have been elevated into positions of authority within local authorities and the police. Vuilliamy (2012) and Nettelfield and Wagner (2013) have discussed the difficulty of talking about reconciliation in the absence of an admission of guilt on the part of the perpetrators, i.e. the impossibility of *reconciliation* without *reckoning*. Summerfield (2003) illustrates this point through a case study of a Bosnian migrant to London who does not respond to any medicalised intervention but whose mental health appears to improve (temporarily) following two (social) episodes related to his home country. Summerfield cites the response of one Bosnian asylum-seeker on being offered assistance by a mental health project: 'We are not mad, we are betrayed' and concludes that it was a shared understanding of the conflict in Bosnia which kept this 'patient' attending therapy sessions for three years. Summerfield argues that instead of attempting to maintain political 'neutrality', a mental health clinician should be probing the 'social rather than the mental space'.[3]

Given the situation in Bosnia where so many who were subject to horrendous violence, rape, torture and forced displacement are still awaiting justice and retribution, it is difficult to hypothesise as to whether positive steps towards a reckoning and reconciliation would result in better mental health outcomes for survivors. Being neither a clinician nor a mental health professional, I am in any case not in a position to offer a medically-informed judgement on the pathology of those migrants from the former Yugoslavia who engaged in my study. However, as a researcher who has been engaged in a professional capacity for years with vulnerable migrants, I was struck by the intensity and the frequency with which my respondents reported their experience of dreams and the act of dreaming and the ways in which the motif of dreams would both puncture and permeate throughout the narratives of their experiences related to migration and to settlement. Particularly striking was the way in which the dream narrative as retold by my respondents differed markedly from the popular discourse on dreams and dreaming associated with migration which emphasises the aspirational nature of migrant (hopes and) dreams.

2015 saw the twentieth anniversary of the end of the war in Bosnia. Anniversaries and commemorative events are reminders of the atrocities committed on the territory of the former Yugoslavia, however the eyes of the world have, in the main, moved in the years since that time to the sites of suffering amongst countless other conflicts across the globe. 2015 and 2016 were also years which saw migration-related events amounting to a real humanitarian crisis playing out across and at the borders of Europe. Such human suffering is a poignant illustration of the ways in which the dream versus reality paradox is a daily nightmare for many and we can only imagine what the current and future night-time dreams of those struggling to seek refuge across Europe today may reflect. The majority of my respondents – regardless of where they come from in the former Yugoslavia – have lived in Britain for more than twenty years. For many, the intensity of their memories and dreams are undiminished over time. Drawing on Summerfield's (2003) social vs mental space argument, a reckoning within the social space may help to bring an emotional reconciliation within the mental space.

Notes

1 'Migrants cling to their dreams as chaos grows in Calais', *ft.com*, 5 September 2014; 'The Maltese dream that's more a migrant's nightmare', *Independent*, 11 August 2013; 'At Greek port, migrants dream and despair in abandoned factories', *Reuters*, 6 May 2015; 'Dream of better life in Britain fuels Calais migrant chaos', *Reuters*, 7 July 2015.
2 Interview recorded for the Evelyn Oldfield Unit Refugee Communities History Project and archived at the Museum of London (reference 2005.119).
3 This does however beg the question as to whether a Bosnian Serb in need of mental health intervention would perhaps benefit from visiting a different therapist.

References

Bajić-Hajduković, I. (2008) 'Belgrade parents and their migrant children', PhD thesis, University College London.

Bal, E. and Willems, R. (2014) 'Introduction: Aspiring migrants, local crises and the imagination of futures "away from home"', *Global Studies in Culture and Power* 21(3): 249–258.

Bogić, M., Ajduković, D., Bremner, S., Francisković, T., Galeazzi, G.M., Kučukalić, A., Lečić-Toševski, D., Morina, N., Popovski, M., Schützwohl, M., Wang, D. and Priebe, S. (2012) 'Factors associated with mental disorders in long-settled war refugees: Refugees from the former Yugoslavia in Germany, Italy and the UK', *The British Journal of Psychiatry* 200: 216–223.

Cartwright, E. (2011) 'Immigrant dreams: Legal pathologies and structural vulnerabilities along the immigration continuum', *Medical Anthropology: Cross-Cultural Studies in Health and Illness* 30(5): 475–495.

Chimienti, M. (2010) 'Selling sex in order to migrate: The end of the migratory dream?', *Journal of Ethnic and Migration Studies* 36(1): 27–45.

Christou, A. (2016) 'American dreams and European nightmares: Experiences and polemics of second-generation Greek-American returning migrants', *Journal of Ethnic and Migration Studies* 32(5): 831–845.

Derluyn, I. and Broekaert, E. (2005) 'On the way to a better future: Belgium as transit country for trafficking and smuggling of unaccompanied minors', *International Migration* 43(4): 31–56.

Du Phuoc Long, P. and Ricard, L. (1996) *The Dream Shattered: Vietnamese Gangs in America*. Boston: Northeastern University Press.

Esson, J. (2013) 'A body and a dream at a vital conjuncture: Ghanaian youth, uncertainty and the allure of football', *Geoforum* 47: 84–92.

Eytan, A., Gex-Fabry, M., Toscani, L., Deroo, L., Loutan, L. and Bouvier, B.A. (2004) 'Determinants of post-conflict symptoms in Albanian Kosovars', *Journal of Nervous and Mental Disease* 192: 664–671.

Fozdar, F. (2009) ' "The Golden Country": Ex-Yugoslav and African refugee experiences of settlement and "Depression" ', *Journal of Ethnic and Migration Studies* 35(8): 1335–1352.

Glick Schiller, N. and Fouron, G.E. (2002) *Georges Woke Up Laughing: Long-Distance Nationalism and The Search for Home*. Durham, DC: Duke University Press.

Halilovich, H. (2013) *Places of Pain: Forced Displacement, Popular Memory and Trans-Local Identities in Bosnian War-Torn Communities*. New York and Oxford: Berghahn.

Horst, C. (2006) 'Connected lives: Somalis in Minneapolis, family responsibilities and the migration dream of relatives'. UNHCR Evaluation and Policy Analysis Unit.

Jančić, M. (1996) *The Flying Bosnian: Poems from Limbo*. London: Hearing Eye.

Johnson, H. and Thompson, A. (2008) 'The development and maintenance of post-traumatic stress disorder (PTSD) in civilian adult survivors of war trauma and torture: A review', *Clin Psychol Rev* 8(1): 36–47.

Knipscheer, J.W., Drogendijk, A., Gülşen, C.H. and Kleber, R.J. (2009) 'Differences and similarities in posttraumatic stress between economic migrants and forced migrants: Acculturation and mental health within a Turkish and a Kurdish sample', *International Journal of Clinical and Health Psychology* 9: 373–391.

Lévy, F. and Lieber, M. (2008) 'Northern Chinese women in Paris: The illegal immigration-prostitution nexus', *Social Science Information* 47(4): 629–642.

Marić, V. (2009) *Bluebird: A Memoir*. London: Granta.

Mollica, R.F., Brooks, R., Tor, S., Lopes-Cardozo, B. and Silove, D. (2013) 'The enduring mental health impact of mass violence: A community comparison study of Cambodian civilians living in Cambodia and Thailand', *Int J Soc Psychiatry* 60(1): 6–20.

Morina, N., Rudari V., Bleichhardt, G. and Prigerson, H.G. (2010) 'Prolonged grief disorder, depression, and posttraumatic stress disorder among bereaved Kosovar civilian war survivors: A preliminary investigation', *Int J Soc Psychiatry* 56(3): 288–297.

Munro, G. and Pritchard, C. (2013) 'Support needs of male victims of human trafficking: Research findings'. The Salvation Army.

Nettelfield, L.J. and Wagner, S. (2013) *Srebrenica in the Aftermath of Genocide*. New York: Cambridge University Press.

Parutis, V. (2011) ' "Economic Migrants" or "Middling Transnationals"? East European migrants' experiences of work in the UK', *International Migration* 52(1): 36–55.

Pillar, G., Malhotra, A. and Lavie, P. (2000) 'Post-traumatic stress disorder and sleep – what a nightmare!', *Sleep Medicine Reviews* 4(2): 183–200.

Poli, R. (2006) 'Migration and trade of African football players: Historic, geographic and cultural aspects', *Africa Spectrum* 41(3): 393–414.

Priebe, S., Janković Gavrilović, J., Bremner, S., Ajduković, D., Franciskovic, T., Galeazzi G.M., Kucukalić, A., Lecic-Tosevski, D., Morina, N., Popovski, M., Schützwohl, M. and Bogić, M. (2013) 'Psychological symptoms as long-term consequences of war experiences', *Psychopathology* 46(1): 45–54.

Priebe, S., Matanov, A., Janković Gavrilović, J., McCrone, P., Ljubotina, D., Knežević, G., Kučukalić, A., Frančišković, T. and Schützwohl, M. (2009) 'Consequences of untreated posttraumatic stress disorder following war in former Yugoslavia: Morbidity, subjective quality of life, and care costs', *Croatian Medical Journal* 50: 465–475.

Silove, D., Momartin, S., Mamane, C. and Manicavasigar, V. (2010) 'Adult separation anxiety disorder among war-affected Bosnian refugees: Comorbidity with PTSD and associations with dimensions of trauma', *Journal of Traumatic Stress* 23(1): 169–172.

Simić, A. (1999) 'Machismo and Cryptomatriarchy: Power, Affect and Authority in the Traditional Yugoslav Family', in S. Ramet and B. Magaš (eds) *Gender Politics in the Western Balkans*. Pennsylvania: Pennsylvania University Press.

Summerfield, D. (2003) 'War, exile, moral knowledge and the limits of psychiatric understanding: A clinical case study of a Bosnian refugee in London', *International Journal of Social Psychiatry* 49: 264–268.

Teo, S.Y. (2011) ' "The Moon Back Home is Brighter"? Return migration and the cultural politics of belonging', *Journal of Ethnic and Migration Studies* 37(5): 805–820.

Teodorescu, D.-S., Siqveland, J., Heir, T., Hauff, E., Wentzel-Larsen, T. and Lien, L. (2012) 'Posttraumatic growth, depressive symptoms, posttraumatic stress symptoms, post-migration stressors and quality of life in multi-traumatized psychiatric outpatients with a refugee background in Norway', *Health and Quality of Life Outcomes* 10: 84. DOI: 10.1186/1477-7525-10-84.

Van Meeteren, M. (2012) 'Transnational activities and aspirations of irregular migrants in Belgium and the Netherlands', *Global Networks* 12(3): 314–332.

Vojvoda, D., Weine, S.M., McGlashan, T., Becker, D.F. and Southwick, S.M. (2008) 'Posttraumatic stress disorder symptoms in Bosnian refugees 3½ years after resettlement', *Journal of Rehabilitation Research & Development* 45(3): 421–426.

Vuilliamy, E. (2012) *The War is Dead, Long Live the War. Bosnia: The Reckoning*. London: Vintage Books.

Wang, C. (2013) 'Place of desire: Skilled migration from mainland China to post-colonial Hong Kong', *Asia Pacific Viewpoint* 54(3): 388–397.

Weine, S.M, Becker, D.E., Vojvoda, D., Hodžic, E., Sawyer, M., Hyman, L., Laub, D. and McGlashan, T.H. (1998) 'Individual change after genocide in Bosnian survivors of "Ethnic Cleansing": Assessing personality dysfunction', *Journal of Traumatic Stress* 11(1): 147–148.

5   Cultural banks and beacons

Portes *et al.* (1999) outline how transnational action can be separated into three different analytical fields: economic, political and sociocultural. Although transnational activities and practices in these three domains can be expected to overlap, Itzigsohn and Saucedo maintain that sociocultural transnationalism should be considered separately from political or economic transnationalism (Itzigsohn and Saucedo, 2002: 768). They define socio-cultural transnationalism as 'transnational practices that recreate a sense of community based on cultural understandings of belonging and mutual obligations' (ibid.: 767). This definition, focussing as it does on the community, appears to be rooted in an understanding of transnationalism as dependent on diasporic connections and, whilst emphasising the institutional level of transnational activities and practices, does not include space for sociocultural transnationalism which could be demonstrated by individuals or even families. Such an understanding of socio-cultural transnational ties is reflective of the shared and collective experience of culture itself (Assman & Czaplicka, 1995).

Through some of the narratives of migrants from the former Yugoslavia we can observe transnational cultural expressions which could be considered highly individualistic in nature. We see for example cultural output connected to the homeland – in terms of literature, art, music – which could be considered to have multiple impetuses. On one level these outputs are (often) in the public domain and as such are available for collective consumption and interpretation and reinterpretation. Those outputs then form a cultural bank to be drawn upon by other members of the diaspora in helping to form and articulate their own particular transnational character. However those producing such cultural output may also be exercising such creativity as a means of responding to their own more individualistic needs of understanding and articulating their relationship with their homeland, the loss of the original 'home' and their new status.

Wilding (2012: 508) discusses how the concept of 'culture' is becoming more disassociated from place and individual groups. Wilding's understanding of cultural expression is particularly relevant to the transnational space where influences and perspectives from different environments can merge and collide leading to the development of the kind of hybrid consciousness as observed by Hannerz (1997). I discuss in this chapter examples of diasporic attempts to create and mould exclusive physical space which I define as 'cultural beacons' – spaces which delineate physical and symbolic diasporic 'territory'. Gupta and Ferguson (1992: 11) have articulated ways in which the diaspora can use collective memory to reconstruct and redefine their new space in the diasporic home, arguing that migrants can 'use the memory of place to construct imaginatively their new lived world'. The construction of physical or metaphorical space can carry particular resonances for those who have experienced the physical destruction or appropriation of their own individual or collective space in the country of origin.

Guarnizo (2003: 669) states 'everyday transnational practices are not easily compartmentalised, nor are their consequences'. Guarnizo was writing in the context of economic transnational activity, arguing that the economic effects of transnational migration are not solely related to activities which could be considered 'economic' in nature. Similarly, we can see here that those activities that at first glance appear to be 'cultural' in nature can also, as is with the case of more overt political lobbying, be vehicles through which some members of the community elite attempt to 'correct' perceived 'misconceptions' of the dominant cultural and wider discourse.

I explore in this chapter how cultural outputs can be conceptualised as: a means through which the individual migrant can articulate and therefore start to process their migration experience and status as a migrant; a way of presenting 'culture' as a focus of celebration and a conduit for change – specifically challenging perceived misconceptions and established understandings; and as 'beacons' of welcome or of warning.

The bank of cultural material

Cultural expressions produced by migrants from the former Yugoslavia in Britain and directed towards and related to the different diasporic collectives are extensive. Across all the diasporic groups we are able to see multiple examples of the production and celebration of book publications, musical works, exhibitions, theatre productions. Some of these are clearly produced by and for the use of the diasporas themselves with others aiming to cross-fertilise into the audience of the non-migrant British population; some will represent individual cultural expressions, others will be group productions involving other members of the diasporic group and some will be produced in collaboration with non-diasporic artists.[1]

Language as a cultural marker has been discussed in the context of individual and collective identity (Schmid, 2001; Barker & Galasiński, 2001). All the diasporic communities from the region in Britain run their own language schools. Usually taking place on a Saturday, these schools are mainly for younger members of the diaspora – the children of migrants who may have been born in the region but most of whom were born to migrant parents in Britain. Many of these courses are funded by the communities themselves, however the Croatian Ministry of Education funds the course in London. The Serbian Society also runs a course 'Serbian for foreigners' for adults. Some of these schools have been operating for a considerable length of time and far from being 'just' spaces for education and training act as focal points where communities congregate and socialise. The parents of younger children for example will often stay on the premises whilst their children are being taught, to drink coffee and chat with the parents of other children. For older children these schools can be viewed as informal 'youth clubs'. One Bosnian interviewee for example met his now wife at one of these schools:

> Most of my friends outside of work are Bosnian, there is a kind of loose informal community, there's no formal community organisation in London as such, well there is but they work mainly with older people who need their help with translation and that kind of thing. For us younger people, we used to hang around the Bosnian schools years ago as children, then we kind of lost touch a bit but now all the different enclaves are coming back together again. Now we tend to congregate around Bosnian events, no formal links exactly, just ad-hoc gatherings.
>
> (Bosnian interviewee, male, 36)

The social aspect of language learning therefore contributes to the community dynamic in very real ways through regular meetings of children, students and parents with common aims of heritage 'reinforcement'.

We can also see ways in which the politics of language may be being used in articulations of identity. Both Kosovar and Serb community groups have been involved in campaigns for the development of GSCE programmes in Albanian and Serbian respectively. These campaigns articulate a desire not only to provide goals for students of the language but also represent attempts to formalise and legitimise aspects of the ethnic identity through 'official' recognition in the eyes of the dominant educative and cultural discourse in Britain.

Some of those interviewed talked about their efforts to encourage language acquisition and language development within their children, efforts which included the use of Saturday language schools, sending children for extended summer holidays with the grandparents and the purchase of books in the 'mother-tongue'. Branko, from Belgrade whose narrative is included in Chapter 1, spoke about how his sister, who was born in England, soon learnt the importance of speaking Serbian at home:

> My sister, and remember she was born in England right? Not like me. My sister would try speaking in English to my mum and my mum would just ignore her. So my sister learnt straight away where her loyalties were expected to lie. She's in her 60s now my sister and she is more Serb I reckon from being here than she ever would have been if she had been born and raised over there. I remember once, she was talking to my mum and she couldn't remember the Serbian word for 'mend', my mum ran crying from the room. My family never relinquished their 'Serbianness', they always strongly upheld the ideology of being 'Serb'.
>
> (Serbian interviewee, male, age 62)

Other interviewees made reference to ambiguous feelings towards encouraging their children to learn the language of their homeland, because of potential associations:

> My kids are still very small, I don't know to be honest about the whole language thing. I don't want my kids to carry all that with them. It's too much, I want it to stop with me. Yeah, ok, a bit of music is alright, a few words, but what goes with all that is just too much.
>
> (Bosnian interviewee, female, 38)

We can see through the words of this Bosnian interviewee how the desire to retain aspects of her heritage was juxtaposed against a need to protect her children from parts of her personal history which she felt they didn't need to learn about. Another interviewee said:

> I know one day, I am going to have to sit down and tell them everything. I don't want them growing up so naïve as to think it's all a bed of roses. But not yet though, it's going to take me a while to think about the best way of telling them everything that happened. Maybe I should write it down?
>
> (Bosnian interviewee, male, 43)

We can see just through some of these articulations how the use of language can carry with it complex and hidden desires to express identity – language as (positive) heritage reinforcement and also language as symbolic of a different kind of personal (negative) heritage – that of violence, the fear of violence, and the forcible destruction of the home environment.

Basch *et al.* writing in 1994 about the transnational condition articulated how migrants attempted to reflect their positioning towards the homeland through their writing: 'It is only in contemporary fiction that this state of "in-betweenness" has been fully voiced' (Basch *et al.*, 1994: 8). King *et al.* (1995: preface) describe the benefits that considering the 'non-academic' literature produced by migrants can give researchers into the experiences of those they seek to understand: 'such insights are often infinitely subtler and more meaningful than studies of migrants which base themselves on cold statistics or on the depersonalised, aggregate responses to questionnaire surveys'. Andrea Pisac argues for a reconceptualization of the home/exile literary experience as a means of developing a better understanding of the daily experience of the writer in exile. One interpretation of this view could be that the act of writing itself becomes a transnational state of being. Pisac (2010: 123): 'The act of writing [...] functions as a device of remaining in touch with the pan-nationalist past and serves as a new cognitive home'. King *et al.* (1995: preface) take the analogy even further: 'For those who come from elsewhere, and cannot go back, perhaps writing becomes a place to live'.

We can see through the work of writers from different parts of the former Yugoslavia in Britain how the writing of the experience of migration can be considered a cultural product in itself forming a bank of cultural references which are subsequently drawn upon, rearticulated and reinterpreted by other members of the diaspora in their own transnational and diasporic imaginings. The two extracts below illustrate how the moment of departure from the homeland is remembered by two Serbian women, the first a writer, the second an interviewee:

> My mother, my father and my sister drove us to the bus station in our white Skoda. It was a strange beginning to a voyage, a weird mixture of holiday and funeral. No one knew what to say. But go I must, and I went. We hugged for what seemed like hours, saying nothing, and I climbed on the bus in the full blast of some mournful southern tune. My parents and my sister stood outside and waved, silent, like creatures in an aquarium. My mother was the smallest of the three [...] As the bus pulled away, she suddenly started running towards it, for no more than five or six yards, and stopped, frozen, just looking towards me.
>
> (Goldsworthy, 2005 writing of a departure in 1986, aged 25)

My parents and sister took me to the bus station. It was one of the most memorable moments of my life and it's as though it's imprinted on my brain which has returned to me in dreams so many times. Nobody spoke, we just held each other, my poor sister, I think she thought I was abandoning her. My parents looked so old and I worried I would never see them again. I looked out of the window of the bus as it left and the look on my mother's face [...] It is so painful even now.

(Serbian interviewee who left Belgrade in 1992, aged 19)

Both the extracts above highlight how the moment of departure (as discussed in Chapter 1) features within the personal narratives of those who have left, even many years later. The second citation also reflects the dream allegory referenced in the previous chapter. The juxtaposition of the two narratives together reinforce the ways in which the writing of the diasporic experience reflects the personal histories of those who have experienced the rupture of family life but could also highlight how exposure to such literature can help form the telling and retelling of the personal narrative.

The act of producing the cultural output in whichever form – written, musical, pictorial – could also be considered a transnational act in itself. The process the artist goes through in the production of the output forces a retelling and a revisiting of the experience in the homeland, the migration experience and the new home. The following extracts are from interviewees who articulate the way they use their writing to cope with what they have experienced:

I wouldn't like to go as far as to say that I found my writing to be a way of dealing with everything that had happened to me. Probably not in an official therapy kind of way. I have never had counselling and am not sure really what a psychologist would say about it all to be honest. I still write, I thought in Bosnian mainly but actually now that you've made me think about it, I think there are large chunks in English. I haven't really sat down and thought about it too much, nobody has really asked me before. But now that I think about it, it would make more sense to write in English if I want the grandkids to actually understand it. Because the chances of them being able to read Bosnian is probably about two percent.

(Bosnian interviewee, female, age 40)

The extract above from the interview with Bosnian migrant, Sara, whose words on Sarajevo roses, introduced this book, shows how her writing has been a conduit through which she channels experiences she deems necessary to 'deal with' in some way, not all directly related to her experiences in the country of origin. The extent to which the writing is 'for her' only or for others is something she reflects on in her thoughts on how her future grandchildren may approach her writing. The potential readership of such writing is also reflected upon by Bosnian migrant, Enis:

> I know it might sound a bit pretentious but it has really made a differ-ence to me. I could even say it saved my life. I felt like I was in some kind of black hole all the time, it was terrible. I started off by writing about superficial stuff, short stories about shopping, barbecues, going to the beach. It was like some kind of anaesthetic numbing of the pain before getting to the abscess. Maybe it will be published after I am gone.
>
> (Bosnian interviewee, male, age 49)

Enis clearly articulated how he saw the act of writing as a means of coping with his experiences. Transnational spaces through these narratives are highly-personalised and individualistic environments, where others are invited almost as an after-thought.

The creation of 'cultural beacons'

The ways that different community groups from the former Yugoslavia organise themselves across Britain include the use of multiple different types of space. Churches and mosques have been established, some with social or community clubs attached and community centres for the exclusive use of the group have been fundraised for and established. Some groups rent space within more 'general use' buildings, outdoor spaces at or for festivals and fêtes are appropriated, charitable organisations operate on a full or part-time basis within dedicated buildings and (usually Saturday) schools have been established to teach the next generation of Croats/Bosnians/Kosovars/Serbs/Slovenes. Cultural information centres have been established, some operating virtually and some long-since closed, cafés and restaurants have been opened and closed; pubs are appropriated for football matches and other sporting events and, in the case of London Serbs, access has been negotiated to space within a local library for Serbian literature.

It is possible to see how the different communities have appropriated whatever space is available to them – limited or extensive – to fit their practical requirements and, often more importantly, their desire and need to express their own 'brand' of cultural identity or symbolic heritage. One Kosovar interviewee described how the mirroring of café culture in cit-ies, towns and villages of Kosovo within areas of Britain where Kosovar Albanians have clustered as residents is physically helping to form 'common ground' on which a more metaphorical understanding between older and newer members of the diaspora is being developed.

> There are some, let's call them 'teething problems' between those who have been here for a longer time and those who are more newly arrived. I have heard some who have been here for a while get upset thinking that the younger ones should have stayed to 'defend the home'. What is happening is that groups have formed in different bars and restaurants and people come and drink coffee and play cards and chat and it really does help. People can see that we have more in common than what divides us.
>
> (Kosovar interviewee, male, age 49)

One Kosovar Albanian café in London in particular has taken on something of an iconic status, especially during the years of struggle for Kosovan independence when politicians and diplomats would carry out discussions on the premises.[2]

The example of café culture above demonstrates how some members of the diaspora can appropriate any physical space available to them which may not be solely dedicated to the use of the diaspora but can be adapted and readapted as and when necessary. Drawing on the work of Massey and of Appadurai, Gielis (2009: 271) shows how transnational migrants' feelings of home can converge around one physical space or place within the new homeland as 'sites where transmigrants can reach out to people in other places'. In the case of some groups from the former Yugoslavia we can see examples of spaces which are exclusively dedicated to the celebration of that particular cultural heritage and which stand as proud symbols of particular traditions associated with the homeland acting as 'cultural beacons' to the diaspora in the new home.

Blunt and Dowling (2006: 2) define the concept of home as a 'spatial imaginary', as a space which can represent 'feelings of belonging, desire and intimacy' but also of 'fear, violence and alienation'. Migrants within the diaspora can both welcome and reject ideas of home, which are to each individual and family highly personal imagined environments, depending on a whole host of factors related to the pre-migration experience and the development of the new home. The destruction and attempted obliteration of sites of cultural heritage was one of the harrowing and distressing features which characterised the wars in Croatia, Bosnia and later in Kosovo. The deliberate attempts to erase those monuments held as symbols of ethnic and national identity and cultural expression – museums, libraries, places of religious worship, bridges – has altered the cultural landscape of some parts of Bosnia and Kosovo with reconstruction of key sites being highly politicised (Walasek *et al.*, 2015). Such deliberate destruction of sites of heritage has left footprints within the collective memory, footprints which have possibly played a part in the development of diasporic space. It is possible to interpret the ways in which the communities in Britain form physical spaces as a response to the actual or perception of threat in the region of origin.

During the course of interviews with community representatives, I would be shown proudly around spaces in which concentrated efforts had been made to reconstruct a 'home from home' in terms of the make-up, atmosphere and character of the physical space. For some this would very much represent a safe space – a beacon of welcome. For others it could signal a different kind of message – a beacon of warning. We can see through the development of these 'cultural beacons' how such sites can be spaces of both 'home from home', places to welcome members of the diaspora and can also be spaces of exclusivity, sites to be avoided for those who do not wish to associate with such symbolic imagery for different reasons. The combination of some sites with the associated ethno-national and/or religious connotations can mean, for some, that these spaces represent familiar but not always welcome associations.

There were a number of respondents who offered explanations as to why they chose not to officially affiliate themselves with a group on a formal basis.[3] Reasons offered included wishing to disengage from the past, not wishing to associate themselves in a formal way with anything they perceived to be 'official' representation and wishing to avoid what they perceived to be an 'ethno-nationalist' focus or a religious identity[4] to which they did not ascribe[5]:

Why don't I want to get involved? Am not sure, have never really felt the need, sometimes I feel as though they all live in the past. We need to move on, that's what I feel like saying to them.

I am a member but only so that I can get to hear about cultural events – I don't gossip and I don't pray so am not a real member.

No, am not a member, it's all barbecues and folk music, that's not my cup of tea.

I question the energy behind some of these organisations, am never sure how much of it is spontaneous and how much is directed by those back home.

Am not religious and they've all gone a bit that way since they've been over here. People who would never have darkened the door of a church back home are suddenly crossing themselves and have gone all devout.

The irony of it is that most of the people here now, who go to mosque and all that, I just don't think they would have done that before, but now it's a bit of a social thing and also, I think a respect thing. There are people here who have their place you know and they need to feel like they're 'someone'.

A lot of them are still living over there mentally. They may be sitting in a hall in London but attitude-wise they're definitely still back over there. I used to be quite active with [name of organisation] but their mentality just gets on my nerves. They [the organisation] are always talking about bringing over some hotshot from [name of city] but they don't realise they're nothing here. They're nobody. They're after hotels for them and taxis and all the fancy stuff, when are they going to realise that nobody here gives a shit about some [name of city] bigwig.

Partly as a consequence of the dispersed nature of migration from the region across the globe, some observers have highlighted how virtual communities have been developed as a means of managing the geographic scattering of the diasporas (Mazzucchelli, 2012), specifically related to the 'Yugoslav' dynamic. Many of the websites which had been fairly active relating to the development of a 'cyber-Yugoslav' space at the beginning of my research had closed down by the time I was writing up the findings of the research.[6] A number of participants in my research demonstrated an ambivalent attitude towards online communities specifically related to the diaspora – 'I don't go on websites, people are always arguing on websites'. Certainly some of those sites reviewed for the purpose of this research would appear to be highly contested sites where the dynamic is one of (in the words of one interviewee) 'fanaticism and furore'. There were also several other examples of less (obviously) politicised online spaces which have been styled as practical advice and support sites.

Culture as a conduit of change

Diasporic groups from the former Yugoslavia in Britain would appear to have thriving spaces of culture with multiple opportunities for sharing in and/or creating cultural expressions. Events such as seminars, exhibitions, concerts, talks, book launches, plays produced by and related to the different diasporic groups are frequent occurrences, especially in London. Some community groups appear to have been established with the express aim of developing relations between the country of origin and Britain and the objective of promoting the culture of the sending society within the host state is explicit within such groups.[7] There are other organisations which also have a more implicit aim of 'enhancing' the understanding of their country of origin or culture within the host society. One representative of a community organisation discussing with another member whether they should agree to my request for an interview was heard to say, 'I think we should say yes, it could be good for us'.[8] Such 'enhancement' of understanding carries with it an element of power and control; those cultural aspects which are for public display will naturally form those which the dominant elite consider to be the 'brightest and the best' that 'their' culture has to offer.

The British-Croatian Society sees culture as a 'subtle' way of promoting Croatia.[9] The British-Croatian Society organised a festival 'Welcome Croatia' with events running throughout 2013 to mark Croatia's accession to the EU with photographic exhibitions, children's theatre performances, an exhibition of Croatian architecture, an archaeology conference, a Croatian national stand at the London Book Fair, concerts and the ringing of bells at a church in North London. In a pamphlet produced to document the artistic links between Britain and Croatia, the author states:

> To find a work of art by one's compatriot in a major world museum is always an exciting moment. It is a link with your homeland and it is also another link with the country you visit or in which you have settled [...] The presence of these works of art show us that links between Britain and Croatia are stronger and more pervasive than is generally known. The movement of artefacts and exchange of ideas has been ongoing for centuries [...] We are proud that a small nation such as Croatia has contributed to the world's rich tapestry of artistic achievements and scientific discoveries.[10]

In October 2011 a series of events, 'Days of Alternative Culture, Art and Science', was run by the Serbian LSE Alumni Association entitled 'Different Face of Serbia'. Including a photography exhibition, a panel discussion on 'Serbia's European challenges', the launch of an interactive online library of Serbian cultural heritage, films, musical performances and other presentations, the series had ambitious aims. The organisers in their 'message from the authors' claimed that 'by focusing on the individual intellectual and artistic excellence and achievements in service of the whole humanity we are eliminating the existing nation-related prejudices and stereotypes'.[11]

One Bosnian migrant to London, who together with his mother, performs *sevdalinke*, traditional Bosnian songs, articulated how he sees the performing and teaching of such music as a fight back against the practice of cultural genocide against Bosnians and against the way in which mass media in Yugoslavia embraced 'turbo-folk'.[12] Through his work, he aims to both keep alive this particularly type of music whilst also educating others of the threat posed to such traditions.

It is evident, especially in those outputs which are targeted at 'others', how the cultural production can be conceptualised as a conduit of change – by aiming to develop a 'better' understanding of the culture in question or to 'correct' apparent misconceptions. Through the formation of cultural references, we can see how migrants both invest in and draw upon 'their' diasporic cultural 'bank' in forming and articulating their sense of individual and collective identity. Cultural symbols and rituals are celebrated within whatever space is made available with some groups having maximised the use of sole dedicated space as cultural 'beacons' for diasporic locals, articulated as a 'home from home'. The transnational physical (and we could assume metaphorical) space is thus intensified receiving a 'boost' from the symbolic imagery and ceremony of the homeland. It is also clear however how the exclusivity of particularly online spaces cannot be relied upon and can be sites of both celebration and contestation. Respondent narratives have demonstrated how the nature of the cultural activity produced is likely to change depending on the target audience and the extent to which the activity is internalised or 'external-facing'. For some the cultural output is a way of expressing transnational emotions to be drawn upon and shared by others both within and outside the diaspora and for others the mechanics of producing the cultural piece can be understood as a transnational act in itself as a way of articulating the migration experience. Considering the ways in which language itself is being used by some members of the diaspora as a cultural identifier and as a focus of the community, it would be interesting to explore further how language and materialism will develop in the future in the descendants of those who have migrated from the former Yugoslavia.

Notes

1 Examples include projects with British photographer Paul Lowe, artist Nicola Atkinson and playwright Sarah Kane.
2 Interview with member of staff at Café Koha, London.
3 See Skrbiš (1999) for a discussion of 'community' membership within the context of Croat and Slovenes in Australia.
4 Expressed affiliation with one of the organised religions amongst my survey respondents was significantly lower than as indicated in census results from different parts of the former Yugoslavia. This discrepancy of religious affiliation between those in the homeland and those in the diaspora could potentially be explained by a combination of cultural expectations in the country of origin, the associated pressure of affiliating with the dominant religious discourse when completing something as politically-loaded as the national census and a desire on the part of some members of the diaspora to distance themselves from the dominant political-cultural narrative.
5 I have not presented demographic information to accompany these quotes as the responses represented all ethnicities and were mixed in terms of gender, age and time of migration.
6 Although it is possible that some of the dedicated websites may have been displaced to other more 'mainstream' social media, for example Facebook.

7 The British–Croatian society is one such example, whose stated objectives are: 'to foster the development of good and friendly relations between the peoples of the United Kingdom and the Republic of Croatia and in particular to promote a better understanding and appreciation of Croatia in Britain'.
8 Extract from fieldwork notes, comments made prior to interview with community representative.
9 Interview with representative from British-Croatian Society.
10 Flora Turner-Vučetić, 'Mapping Croatia in United Kingdom Collections', British Croatian Society publication published in conjunction with the 'Welcome Croatia Festival' (series of events to mark Croatia's entry to the European Union).
11 Source: 'Different face of Serbia' promotional leaflet.
12 See Collin (2001) and Baker (2006 and 2007) for discussion of turbo-folk.

References

Assmann, J. and Czaplicka, J. (1995) 'Collective memory and cultural identity', *New German Critique* 125–133.

Baker, C. (2006) 'The politics of performance: Transnationalism and its limits in former Yugoslav popular music, 1999–2004', *Ethnopolitics* 5(3): 275–293.

Baker, C. (2007) 'The concept of Turbofolk in Croatia: Inclusion/exclusion in the construction of national musical identity': http://eprints.soton.ac.uk/66293/1/ Baker_-_turbofolk_2007.pdf (accessed 16 July 2016).

Barker, C. and Galasiński, D. (2001) *Cultural Studies and Discourse Analysis: A Dialogue on Language and Identity*. London: SAGE.

Basch, L., Glick Schiller, N. and Szanton-Blanc, C. (1994) *Nations Unbound: Transnational Projects, Postcolonial Predicaments, and Deterritorialized Nation-States*. New York: Gordon and Breach.

Blunt, A. and Dowling, R. (2006) *Home: Key Ideas in Geography*. Oxon: Routledge.

Collin, M. (2001) *This is Serbia Calling: Rock'n'Roll Radio and Belgrade's Underground Resistance*. London: Serpent's Tail.

Gielis, R. (2009) 'A global sense of migrant places: Towards a place perspective in the study of migrant transnationalism', *Global Networks* 9(2): 271–287.

Guarnizo, L.E. (2003) 'The economics of transnational living', *International Migration Review* 27(3): 666–699.

Gupta, A. and Ferguson, J. (1992) 'Beyond "Culture": Space, identity, and the politics of difference', *Cultural Anthropology* 7(1): 6–23.

Hannerz, U. (1997) 'Flows, boundaries and hybrids: Keywords in transnational anthropology', *Working Paper Series* for the ESRC Transnational Communities Programme, Oxford University, WPTC - 2K – 02. First published in Portugese as 'Fluxos, fronteiras, hibridos: palavras-chave da antropologia transnacional', Mana (Rio de Janeiro) 3(1): 7–39, 1997.

Itzigsohn, J. and Saucedo, G. (2002) 'Immigrant incorporation and sociocultural transnationalism', *International Migration Review* 36(3): 766–798.

King, R., Connell, J. and White, P. (eds) (1995) *Writing Across Worlds: Literature and Migration*. London: Routledge.

Mazzucchelli, F. (2012) 'What remains of Yugoslavia? From the geopolitical space of Yugoslavia to the virtual space of the Web Yugosphere', *Social Science Information* 51: 631–648.

Pisac, A. (2010) 'Trusted tales: Creating authenticity in literary representations from ex-Yugoslavia', PhD thesis, Goldsmiths College, London.

Portes, A., Guarnizo, L.E. and Landolt, P. (1999) 'The study of transnationalism: Pitfalls and promises of an emergent research field', *Ethnic and Racial Studies* 22(2): 217–237.

Schmid, C.L. (2001) *The Politics of Language: Conflict, Identity, and Cultural Pluralism in Comparative Perspective.* Oxford: Oxford University Press.

Skrbiš, Z. (1999) *Long-Distance Nationalism: Diasporas, Homelands and Identities.* Hants: Ashgate.

Walasek, H., Carlton, R., Hadžimuhamedović, A., Perry, V. and Wik, T. (2015) *Bosnia and the Destruction of Cultural Heritage.* Surrey: Ashgate.

Wilding, R. (2012) 'Mediating culture in transnational spaces: An example of young people from refugee backgrounds', *Continuum: Journal of Media & Cultural Studies* 26(3): 501–511.

Conclusion

This study has demonstrated how migrants from the former Yugoslavia have made their decisions to leave their homes and move to Britain at different points in history, many exhibiting a multitude of motivations for making their journeys. The wide variety of migration stories articulated by participants in this project highlight the danger of making assumptions about the migration experience based on country of origin or ethnicity. Scholars of migration (Turton, 2003) are critical of literature which refers to the 'refugee experience' (as a singular unit of analysis); the experiences of those from the former Yugoslavia have shown how difficult it would be to broadly conceptualise the experiences of migrants from the region as one collective experience of migration or transnationalism, by area of origin, by ethnicity or by migration 'status'. The discourse around migration has been and continues to be highly-politicised within all contexts. As the rhetoric intensifies, the voices of those who have made the decision to migrate, with whatever motivation(s), risk becoming engulfed in a storm of politics, theories, competing agendas and efforts to classify 'type' of migration. In problematising the question of migration, the individualisation of anyone who has moved to live – either temporarily or permanently – in a country other than the one in which he/she were born, can sometimes become lost amongst policy responses to the collective and blurred by the academic desire to understand and conceptualise.

This project began life in a desire to consider the relationship that someone who has made his/her home in a different country to the one in which they were born develops with the homeland. The study aimed to explore the dynamic within transnational and diasporic space(s) inhabited by migrants from the former Yugoslavia and the variables which may impact upon the formation of such spaces. The case of the former Yugoslavia adds an important new dimension to any possible debates on transnationalism through the ways in which borders and territories have been drawn and redrawn allowing for potentially multiple and conflicting affiliations and related identities.

The narratives of respondents have demonstrated how these multi-layered affiliations can translate into transnational acts and behaviours and also how such dynamics can then play out within the diaspora.

The research has highlighted the difficulties in categorising migrants from the region by 'type' of migrant along the lines of 'voluntary' or 'forced'; 'economic' versus 'refugee', etc. Capturing such experiences would also be difficult to do on any 'spectrum' of the migration experience, no matter how loose and flexible the framing. Such a spectrum would have to take into account not only the original 'reason(s)' for making the initial decision to leave, for going through the process of packing, deciding what to take and what to leave behind, goodbyes, the fracture of leaving behind the home, work, study, family, friends, the familiar but would also need to take into account motivations for re-enacting that initial departure potentially hundreds of times over the course of a lifetime. It would be difficult to capture on any spectrum the decision-making process over whether/when to return and the ways in which migrants cope with any potential merging or clashing of the current life with echoes of a life pre-migration.

The roots of the literature on transnationalism have principally been grounded in the experiences of labour or voluntary migrants from Latin America to the US. This study has contributed towards the discussion within a European context of those who migrated with multiple motivations. Many of those migrants will also have had the sometimes unfortunate experience of the maze that represents the British asylum system. This research has therefore made a contribution towards addressing the identified gap in the literature around the transnational experiences of those who have not migrated voluntarily. The research has also highlighted how the legal status of a migrant may not always correlate with how the migrant him/herself articulates his/her migration-related identity. For those wishing to migrate under the conditions imposed by the British immigration authorities, some migrants may have felt as though they had little choice but to take whatever status was offered if it meant being able to stay in the country. In many cases the decision-making process on the part of those authorities responsible for bestowing the desired status appears to both the migrant and the outside observer as at best opaque and can at times seem totally bewildering.

Diaspora and transnationalism are both useful concepts through which to view the often complex relationships of migrant – homeland – new home – other migrants – non-migrants. I outlined at the beginning of the book how the diasporic experience could be viewed differently from the transnational in that membership of the diaspora ultimately focuses on the collective, shared experience of those who subscribe. In that sense, the diasporic space can be viewed as one which can cross borders of multiple nation states but is above all a shared space which migrants can enter or leave fluidly depending on a number of different contexts. The transnational space is also a fluid ever-changing space but one which could also reflect and be characterised by highly-individual and personal narratives related to the home country, the new homeland and others in the diaspora. Contrary to the emphasis on the physical return of migrants and the intensity of quantifiable cross-border activity as a condition of transnationality as argued by Portes, I argue for a more individualistic understanding of the transnational paradigm, more in line with the 'transnationalism as a type of consciousness' approach as conceptualised by Vertovec (1999: 447). The study has provided evidence of the experience of those migrants who may be active members of diasporic community groups and has also captured the voices of those who have made a conscious decision to avoid membership of any collective identity and who, yet, retain deep and complex emotional ties to the homeland. A twin diasporic-transnational lens of analysis allows, therefore, for the articulation of collective, individual, shared and very private narratives related to the migration experience.

The conflicts in the former Yugoslavia have been the subject of much intense debate and discussion on the part of politicians, academics and journalists around the catalysts behind the wars and the main culpable party(ies). Such discussions and debates can be drawn upon, argued against, discarded as not relevant or enthusiastically embraced by migrants from the region (and indeed were referenced by participants in this project) depending on the extent to which such discussions reflect the migrant's own experience of the conflict – either through direct personal experience or the collective narrative of the conflict formed by the migrant. The availability of organised community groups either in reality or virtually through online groups, can help to strengthen and reinforce the diasporic narrative. We can see evidence of this in the ways in which a group narrative of history and migration can be shared across communities in the shaping of personal testimonies. Serbian participants in the project in particular made reference to how they 'grew strength' from the experience of others in opposing what they perceived to be unjust accusations of blame. Some members of the Serb community in Britain were amongst the most vociferous and proactive in their attempts to develop positive images of Serbhood to counter what they perceived to be a popular narrative of Serb vilification. The historic relationship between Britain and Serbia at the level of elite 'sponsorship' could have provided a vehicle through which such 'PR'-type activity has been given the opportunity to develop and flourish. Indeed, the longer history of migration from Serbia to Britain and the longevity of the existence of Serb community groups in Britain compared to other communities from the former Yugoslavia, combined with the historic respect afforded to Serbia by British parliamentarians may have contributed to an even greater collective hurt felt by those Serb communities in Britain when that hand of friendship seemed to be withdrawn towards the end of the conflict on Bosnian territory and then later in Kosovo. Such friendship is referred to by representatives of communities from other parts of the former Yugoslavia to partially explain how Serbian aggression in Croatia and the genocidal practice of 'ethnic cleansing' on the part of Bosnian Serbs paramilitaries later in Bosnia was minimised in the popular reporting of those conflicts in Britain. On one level therefore, the ambitious activities of the more proactive members of the Serb communities in Britain today could be indicative of individual personal career aspirations related to their eventual planned return to Serbia but could also be a reflection of a more collective ambition and aspiration: the restoration of historic Serbo-British friendship ties and the 'correction' of what they consider to be the misconception of Serbs as the instigator of Balkan suffering in the 1990s.

Despite the challenges encountered in conceptualising motivation behind migration from the former Yugoslavia along linear and binary spectrums, it is still possible to identify ways in which both such motivations and immigration status in the country of destination can have an impact upon the migrant's relationship with the sending and receiving states and ultimately on the development of any transnational affiliations. Previous literature on the transnationalism has tended to stress the lack of capacity on the part of those who could be considered to be forced migrants to be active transnational participants. But I would argue that any lack of capacity as related to the experience of migrants from the former Yugoslavia cannot solely be attributed to economic or financial limitations of individual migrants, nor is such a lack of capacity a characteristic of only those who could be termed 'forced' migrants. The problems posed by poor financial infrastructure, the homeland's status as international protectorate and perceived levels of corruption can act in conflict with the expressed desire on the part of some migrants to actively engage in a political or economic way with their country of origin.

The 'desire versus capacity' conceptualisations of transnationalism (Al-Ali *et al.*, 2001) and 'resource-dependent' transnationalisms (Itzigsohn and Saucedo, 2002) is also interesting to explore within this context. Discussions of restrictions on contributions to the country of origin have usually been within the context of economic resources. I would suggest however that 'resources', far from referring to only financial means can also be interpreted as encompassing emotional and mental capacity, will and aspiration to engage in anything approaching a transnational or diasporic space. Emotional resources towards the homeland can ebb and flow depending on a whole host of variables, including the context to the migration experience, the age and gender of the individual migrant, the socio-political situation in the country of origin and within the 'new home', the presence or absence of family or friends in the homeland, the position of the individual on the lifecycle and in response to certain catalysts or 'triggers'. The experiences of those from the former Yugoslavia have also borne witness to how the immigration system itself can play a very active role in the physicality of any transnationalism. Any emphasis on quantifying the frequency of visits 'home' as a determiner of transnationalism becomes redundant if, by being expected to renounce any travel documents, the migrant is physically prevented from leaving the country during a protracted determination process of immigration status.

Many of those who participated in this project who came to Britain irrespective of their motivations for migrating report that they had planned to return to the homeland 'when conditions were right'. This is still the case for some who follow events in the country of origin closely, keeping one eye on 'what's happening over there', waiting for the most opportune moment to go back. Some of this 'what's happening over there' relates to the wider socio-political context within the country of origin and for others, who may be looking to set up business or make a living, they are continuously making an informed assessment of the relevant economic conditions. One interviewee in her role working with refugees from the region in London reported that some refugees had been 'unable to cope with their new status in England' and many had returned to Bosnia after the war had ended. Others interviewed talked about 'keeping their options open' and after years of feeling as though they had little or no control over their own lives were demonstrably revelling in having a choice at all.

For those who had considered whether to contribute financially or through political action to the country of origin, the question often posed in determining whether to take such action or not was whether it 'would make a difference'. The 'would it make a difference' or the employment of an individualistic, informal assessment of effectiveness before making the decision as to whether to act transnationally – be it in economic, political or socio-cultural sense – can be seen through the narratives of those migrants who relate their aspiration to 'give something back' to, in this context, the country of origin. The effectiveness test however will often translate into action aimed at the individual or local community level as opposed to national interests, so often perceived to be either under the influence of corrupt and/or international stakeholders and/or not addressing the concerns of most relevance to the individual migrant. The employment of the 'would it make a difference test' before making the commitment to transnational economic and political acts is articulated through the narratives of respondents, irrespective of country of origin or ethnicity, gender or age and could go some way to explaining the different expressions of transnational acts and behaviour. Individual migrants are acting in a way that they believe will have the greatest impact, this impact will be assessed differently depending on variables such as motivations for migrating, perception of the local political and economic contexts in the area of origin, presence of family or friends in the area of origin, perceived levels of need in the area of origin and plans for return.

I would suggest that an understanding of the effectiveness assessment as carried out by transnational migrants would go some way towards providing a response to those who are looking for 'agency' in the study of transnationalism (Baas, 2010) and may also be a way of conceptualising the experiences of those who, having had certain decisions about key aspects of their lives enforced upon them through the experience of migration both in the country of origin and the country of destination, are reasserting control and empowerment. Carling makes a case for an understanding of the transnationalism paradigm through the lens of transnational *lives* rather than transnational *people*.[1] Whilst, I acknowledge that it is important to consider the grounded reality of daily experiences and that labelling individuals as transnational migrants or not risks certain narrow understandings of the transnational paradigms, I would also suggest that at certain points migrants can, by very strongly articulating their transnational affiliations (or absence thereof), identify as (non)transnational individuals for very particular and personal reasons – not least as a reaction to their experiences of the migration process, be they positive or negative.

Much of the literature on transnationalism measures the 'strength' of transnational ties through quantitatively measurable economic and political ties and frequency of visits home (Guarnizo, 2003; Portes *et al.*, 2002). Articulations of transnational activity expressed during the course of the research would be consistent with studies which have found intense and regular political and economic transnational ties to be the exception rather than the norm in the lives of the majority of migrants (Levitt *et al.*, 2003). Such an understanding of transnationalism, emphasising the 'strength' of economic and political ties by implication suggests that those migrants who do not exhibit such ties are expressing 'weak' transnational relationships. The experiences of my respondents would suggest however that a migrant who may have few or no economic or political ties to the homeland but whose daily life is coloured by dreams, imaginings and memories of that homeland exhibits a quality of transnational relationship which could be described as anything but 'weak'.

The experiences of my respondents have demonstrated how some migrants articulate the strength of their transnational ties not through measurable economic or political acts but through complex and multi-layered emotional relationships with scattered family members, with the homeland as a personalised actor in the transnational dynamic and with the imaginings of lives past or lives that could have been. I would strongly advocate for an understanding of the transnational experience (as far as it is possible to speak of such an experience in the singular) that hears the voices of those who do not remit, do not vote, do not own property in the country of origin, have no business interests, are not the chair of their local community organisation and may not even make regular visits 'home' but who carry intense and highly-personalised echoes of the homeland. Through the narratives of those migrants whose lives do not feature elements of political or economic transnationalism, the transnational space is transformed into one not of entrepreneurship and empowerment but a space which is overshadowed by memory and longing – memories of a different life and an articulated desire or imagining for a life that the migration experience has knocked off its trajectory. The multitude of experiences of migration that have been articulated through this study therefore translate into lived experiences of transnationalism which cover degrees of quantity and quality of experience. Through focussing on the varied experiences of those who migrated with different motivations and for different reasons, it is possible to make the case for an understanding of the transnational paradigm which encompasses more nuanced relationships with the homeland that would be immediately suggested by, for example, measuring the remittance count.

Ambivalences have been attributed towards the concept of transnationalism (Harney and Baldassar, 2007: 190) through its conceptual removal from 'the materiality of everyday life, the more local and particular connections and social relations of the people, places, discourses and practices we study'. Ultimately, it is hoped that the articulations of the lived realities of migrants as expressed throughout this book have made a contribution towards narrowing the perceived gap between theories and conceptualisations and the grounded experiences of those on which such theory is based.

Note

1 Jorgen Carling, 'The transnational theatre: Conceptualisations of transnationalism in migration research', RGS-IBG Annual Conference, 1–3 September 2010.

References

Al-Ali, N., Black, R. and Koser, K. (2001) 'Refugees and transnationalism: The experience of Bosnians and Eritreans in Europe', *Journal of Ethnic and Migration Studies* 27(4): 615–634.

Baas, M. (2010) *Imagined Mobility: Migration and Transnationalism among Indian Students in Australia*. London: Anthem Press.

Guarnizo, L.E. (2003) 'The economics of transnational living', *International Migration Review* 27(3): 666–699.

Harney, D.M. and Baldassar, L. (2007) 'Tracking transnationalism: Migrancy and its futures', *Journal of Ethnic and Migration Studies* 33(2): 189–198.

Itzigsohn, J. and Saucedo, G. (2002) 'Immigrant incorporation and sociocultural transnationalism', *International Migration Review* 36(3): 766–798.

Levitt, P., DeWind, J. and Vertovec, S. (2003) 'International perspectives on transnational migration: An introduction', *International Migration Review* 37(3): 556–575.

Portes, A., Guarnizo, L.E. and Landolt, P. (1999) 'The study of transnationalism: Pitfalls and promises of an emergent research field', *Ethnic and Racial Studies* 22(2): 217–237.

Portes, A., Guarnizo, L. E. and Haller, W.J. (2002) 'Transnational entrepreneurs: An alternative form of immigrant economic adaptation', *American Sociological Review* 67(2): 278–298.

Turton, D. (2003) 'Conceptualising forced migration', RSC working paper No. 12, Queen Elizabeth House, University of Oxford.

Vertovec, S. (1999) 'Conceiving and researching transnationalism', *Ethnic and Racial Studies* 22(2): 447–462.

Appendix
Note on methodology

I briefly outline here the methods used during the research process, provide some general demographic information about my respondents and on the methods used for participant recruitment. Throughout the study, I aimed to reflect through the methodology an approach which could be considered to mirror the transnational paradigm itself, to develop holistic tactics which aim to capture as far as possible the transnational character and dynamic of migrants from the former Yugoslavia to Britain.

The basis of the methodology used throughout the study was essentially a mixed-methods approach. I employed some quantitative analysis of data through the administering of the survey (179 responses), and in particular the use of closed questions, but the project was mainly qualitative in nature through: interviews (46) with individual migrants and community representatives (some speaking in a personal as well as an 'official' capacity); a review of print media outputs and parliamentary transcripts; analysis of online discourse by and between diasporic 'community' groups; review of autobiographical contemporary fiction produced by writers from the former Yugoslavia who have made their homes in Britain; ethnographic observation at events run by community groups or related to the former Yugoslavia; and archive work at the Museum of London, the Imperial War Museum and the School of Slavonic and East European Studies.

I identified possible survey respondents and interviewees mainly through a snowball sampling technique (Vogt, 1999 cited in Atkinson and Flint, 2001). One of the often-cited problems in snowball sampling is the issue of selection bias (Groger et al., 1999; Magnani et al., 2005) and the impact that this can have on the findings of research that has only 'sampled' those respondents which have been 'hit by the snowball'. I was conscious from the start of the fieldwork of the possible biases that can arise from snowball sampling, especially if the original snowballs are launched from only limited 'sites'. I also wanted to capture the experiences of those who may not actively identify as 'of the diaspora'. I made a concerted effort therefore to instigate snowballs from as many and as varied locations and pathways as possible, including:

• Postings with information about the study on targeted websites and email lists aimed at the diaspora.
• Distribution of a flyer in English and in translation advertising the study and including a link to the survey on noticeboards and in the foyers of a number of 'key' locations including community centres, language schools, targeted charity shops, libraries, churches and within newsagents and post offices in areas of London where it is known that significant numbers of individuals/families from the former Yugoslavia reside.
• A call for participants was published in *Haber* (March–April 2010), the magazine produced by the Bosnia-Herzegovina UK Network and in *Britić*, a Serbian diaspora publication.
• Targeted mailing of community organisations.
• Targeted mailing of groups, such as wider refugee and migrant support groups, who may have clients from the former Yugoslavia.
• Researcher attendance at community events.
• Personal contacts.

In total I received 203 responses to the survey, 179 of which were useable for analysis. The majority of respondents were based in London, which reflects census data on areas of residence. Other parts of the country where respondents were based included Oxford, Beckenham, Birmingham, Brighton, Cambridge, Canterbury, Cardiff, Edinburgh, Glasgow, Guildford, Leeds, Leicester, Oldbury, Reading, Rochdale, Stockport, Surrey, Wiltshire. The overall gender split of the total sample group was 52 per cent female and 48 per cent male. The respondent group as a whole was highly educated with a large proportion reporting a postgraduate level of education. By country of origin, participants in the project were from Serbia (24%); Croatia (22%); Bosnia (20%); Kosovo (7%); Slovenia (6%). Approximately 20 per cent of respondents stated their country of origin as 'Yugoslavia'; this in some cases was indicative of the time that the individual had left the country of origin (prior to the dissolution of the SFRY) but in the cases of those who had left later was more reflective of an identification with Yugoslavia (and whatever that may represent) over and above any individual constituent part. The framing of the study as being of those from the former Yugoslavia may have resulted in the relatively low response rate from Kosovans. The age of participants ranged from 18–76 with the majority of respondents being in the 30–49 age bracket. Participants from Slovenia and Croatia in particular were of a younger demographic, which would be consistent with the data provided by the Slovenian Embassy and through community representative interviews.[1]

In 'allowing' the data collected through fieldwork to determine the framing of certain aspects of the project, in particular, the focus of the 'context of reception', my approach could be considered to follow theories of constructivist grounded methods as developed by Glaser and Strauss (1967), Glaser, (1992) and then Charmaz (2003, 2005, 2006). Crompton (1998) and Marshall (1989) make the case for the use of methodological pluralism which refers to the 'adoption of methods which enable the researcher to use different techniques to get access to different facets of the same social phenomenon' (Olsen, 2004: 6). By adopting a methodologically pluralistic lens on the project, I aimed to capture as much of the totality of the experience as is possible from a research perspective. I thereby attempted to gain a more nuanced understanding of the 'story' of migration – from the perspective of both the individual and the collective – and to illustrate how such a depth of understanding can help to make a case for an interpretation of transnationalism not just through the 'global' but through the lived experience of migrant individuals and diasporic groups.

Note

1 One Croatian community representative, at interview, stated that the majority of Croats in the UK are here as students, taking advantage of the less expensive education opportunities than in the US.

References

Atkinson, R. and Flint, J. (2001) 'Accessing hidden and hard-to-reach populations: Snowball research strategies', *Social Research Update* Issue 33.

Charmaz, K. (2003) 'Grounded theory', in J.A. Smith (ed.) *Qualitative Psychology: A Practical Guide to Research Methods*. London: SAGE.

Charmaz, K. (2005) 'Grounded theory: Objectivist and constructivist methods', in N.K. Denzin and Y.S. Lincoln (eds) *The SAGE Handbook of Qualitative Research*. 2nd edn. Thousand Oaks, CA: SAGE.

Charmaz, K. (2006) *Constructing Grounded Theory*. London: SAGE.

Crompton, R. (1998) *Class and Stratification: An Introduction to Current Debates*. Cambridge: Polity.

Glaser, B. (1992) *Emergence vs Forcing: Basics of Grounded Theory Analysis*. Mill Valley, CA: Sociology Press.

Glaser, B. and Strauss, A. (1967) *The Discovery of Grounded Theory: Strategies for Qualitative Research*. Chicago, IL: Aldine.

Groger, L., Mayberry, P. and Straker, J. (1999) 'What we didn't learn because of who would not talk to us', *Qualitative Health Research* 9(6): 829–835.

Magnani, R., Sabin, K., Saidel, T. and Heckathorn, D.D. (2005) 'Sampling hard to reach and hidden populations for HIV surveillance', *AIDS* 19(2): S67–S72.

Marshall, G. (1989) *Social Class in Modern Britain*. London: Unwin Hyman.

Olsen, W. (2004) 'Triangulation in social research: Qualitative and quantitative methods can really be mixed', in M. Holborn and M. Haralambos (eds) *Developments in Sociology*. Ormskirk: Causeway Press.

Vogt, W.P. (1999) *Dictionary of Statistics and Methodology: A Nontechnical Guide for the Social Sciences*. London: SAGE.

Index

'ancient hatred' hypothesis 40, 48
arrival and reception, contexts of
    31–52, 117; arrivals in Britain
    32–4; elite relationships 39–43;
    'even Slovenia' theme 47; fear
    of being arrested as 'dissidents'
    52; international 'burden-sharing'
    agreements 34; interview with
    Bosnian migrant 47; interview
    with Croatian community
    representative 47; interview with
    Serbian migrant 47–51; lobbying
    43–5; media discourses 46–7;
    negotiating the British immigration
    system 34–8; 'pan-Slavic' stance
    of comradeship 40; political
    lobbying 50; representations of
    the former Yugoslavia and its
    migrating populations 38–47; royal
    relationships 45–6; transnational
    'triggers' 51–2; virtual call to
    arms 44

'Bosnia Project' 34
'burden-sharing' agreements,
    international 34

contexts of arrival and reception *see*
    arrival and reception, contexts of
contexts of departure *see* departure,
    contexts of
cultural banks and beacons 93–104;
    bank of cultural material 94–9;
    conduit of change, culture as
    102–4; creation of 'cultural beacons'
    99–102; heritage 'reinforcement' 95;

individualistic transnational cultural
    expressions 93; language, politics of
    95; political lobbying 94
cultural identifier, language as 104
cultural identity 99

departure, contexts of 13–26;
    'economic migrants' 19; 'ethnic
    cleansing' 18; historical antagonisms
    15; migration from the former
    Yugoslavia 13–18; motivations
    behind the migration decision 18–25;
    nationalistic fervour 16; referendum
    on independence 17; reflecting on
    the experience of departure(s) 25–6;
    ten-day war 17
diasporic actors 5
'diasporic borrowing' 65–7
dreams of migrants 79; *see also*
    intangible transnationalisms

Early Day Motion (EDM) 42, 44
'economic migrants' 19
elite relationships 39–43
emotions, 'bank' of 58
empathetic experience, accumulation
    of 67
'ethnic cleansing' 42; Bosnian Serb
    forces 18, 110; homes changed
    significantly by 2; Kosovo's
    independence following 17; shared
    collective memory and 6
European Union (EU) 17, 47
European Voluntary Worker
    (EVW) 33
'even Slovenia' theme 47

Federal Republic of Yugoslavia (FRY) 17
forced migrants 110

heritage: language as symbol of 97;
'reinforcement' 95

identity: articulated 60, 64, 95;
avoidance 70–2; collective 104;
cultural, brand of 99; ethnic,
strengthening of 51; formations 9,
14; individual, positioning of 32;
'international' 72; language and 97;
religious 101; shared sense of 6
'in-betweenness,' condition of 4
independence: Kosovan, struggle for
100; referendum on 17
information, communications and
technology (ICT) 2
intangible transnationalisms 79–90;
dream allegory, representation of in
migration narrative 80–1; 'invisible
injuries of soul' 88; juxtaposition of
'there/then' and 'here/now' 82–90;
mental health focus of the refugee
experience 81–2; post-traumatic
stress disorder 81
international 'burden-sharing'
agreements 34
International Criminal Tribunal for the
former Yugoslavia (ICTY) 18, 49
'international' identity 72
'invisible injuries of soul' 88

Kosovo Liberation Army (KLA) 17

language: ambiguous feelings towards
96; cultural identifier 104; emotional
58; identity and 97; politics of 95;
related to motivation for migration 3,
57; as symbol of heritage 97
lexicon of migration experience 57–76;
articulations of migration, exile and
'refugeehood' 59–65; community
absorption and 'diasporic borrowing'
65–7; connotations 57; disloyal
deserter, refugee as 61–2; emotions,
'bank' of 58; empathetic experience,
accumulation of 67; identifications
and identity avoidance 67–72; 'international' identity 72; narrating guilt

and the migration experience 72–6;
peer support 67; 'value-free' term 57
lobbying 43–5, 94
London School of Economics (LSE)
69, 103

Member of Parliament (MP) 41, 42, 68
methodology: data collected through
fieldwork 117; interviews 115;
methodological pluralism 117;
mixed-methods study 115; survey
respondents and interviewees 115

Non-Governmental Organisation
(NGO) 62
North Atlantic Treaty Organisation
(NATO) 17, 40, 51, 72

peer support 67
political lobbying 50, 94
politics of language 95
post-traumatic stress disorder (PTSD)
81, 82
public relations (PR) 44, 109

religious identity 101
royal relationships 45–6

'sampling on the dependent variable' 9
Serbian Relief Fund (SRF) 15, 33
Socialist Federal Republic of
Yugoslavia (SFRY) 16, 66, 116
transnationalism, diaspora and migrants
(conclusion) 107–13; 'ethnic
cleansing' 110; forced migrants 110
transnationalism, diaspora and migrants
(introduction to) 1–9; cautions
against focussing on ethnicity 9;
conceptual understandings 3–6;
diasporic actors 5; experience of
migration 2; 'in-betweenness,'
condition of 4; language associated
with migration 3; literature 8;
research design and chapter structure
7–9; 'roses' 1; 'sampling on the
dependent variable' 9; shared
collective memory 6; terminology
6–7; transmigrants 4; 'triadic
relationship' 5

transnational 'triggers' 51–2
'triadic relationship' 5

United Nations (UN) 17, 21, 51
United Nations High Commissioner for
    Refugees (UNHCR) 22, 31, 48

'value-free' term 57
Vojska Republike Srspke (Bosnian
    Serb Army) (VRS) 49

World War One (WWI) 1, 16
World War Two (WWII) 1, 15, 17, 50